Advances in Anatomy, Embryology and Cell Biology publishes critical reviews and state-of-the-art surveys on all aspects of anatomy and of developmental, cellular and molecular biology, with a special emphasis on biomedical and translational topics.

The series publishes volumes in two different formats:

- Contributed volumes, each collecting 5 to 15 focused reviews written by leading experts
- Single-authored or multi-authored monographs, providing a comprehensive overview of their topic of research

221
Advances in Anatomy, Embryology and Cell Biology

More information about this series at
http://www.springer.com/series/102

Frederic Shapiro

Disordered Vertebral and Rib Morphology in Pudgy Mice

Structural Relationships to Human Congenital Scoliosis

Frederic Shapiro
Laboratory for the Study of Skeletal Disorders
Harvard Medical School
Boston, Massachusetts
USA

ISSN 0301-5556 ISSN 2192-7065 (electronic)
Advances in Anatomy, Embryology and Cell Biology
ISBN 978-3-319-43149-9 ISBN 978-3-319-43151-2 (eBook)
DOI 10.1007/978-3-319-43151-2

Library of Congress Control Number: 2016950901

Printed on acid-free paper

This Springer imprint is published by Springer Nature
The registered company is Springer International Publishing AG Switzerland

Abstract

Disordered vertebral and rib morphology in the pudgy mouse, due to mutations in the *delta-like 3* (*Dll3*) gene of the Notch family, has been studied using whole mount preparations, specimen radiographs, histology, and computerized three-dimensional reconstructions of serial histologic sections. Studies were done in 68 mice, 37 pudgy (pu/pu) and 31 non-affected (pu/+), from late embryo to 3 months of age. *Vertebral bodies* showed malformations including: hemivertebrae; wedged vertebrae; bifid vertebrae (butterfly, double hemivertebrae at same level); and fused vertebrae (partial unilateral bony bar or complete bone block). *Intervertebral discs* adjacent to malformed vertebrae were always abnormal regarding: size (smaller or larger than normal); position (oblique or vertical); shape (round, hexagonal, curvilinear, and other patterns); disorganized peripheral fibers of annulus fibrosus; partial presence across width of vertebral column; and/or complete absence between adjacent vertebrae. *Rib* malformations included: decreased numbers; asymmetric shape and spacing; fused and widened ribs; branching from fused origins into two, three, or more separate ribs; and paravertebral continuous longitudinal cartilage accumulations from which fused ribs originated. The vertebral, intervertebral disc, and rib abnormalities were different in each pudgy mouse. This variability indicates that a single gene mutation does not account solely for the deformities making it necessary to explain by other concepts such as epigenesis. Mutations of genes involved in the oscillatory clock/segmentation mechanism (delta-ll3, Mesp1, lunatic fringe, and others), expressed as early as the presomitic mesoderm stage of embryogenesis, have been clearly implicated in congenital vertebral and rib malformation in mice and humans. Invariable intervertebral disc abnormalities adjacent to abnormal vertebrae are observed, leading to consideration of additional pathogenetic mechanisms. Normal and abnormal structures of vertebral bodies appear directly related to shape and position of the intervertebral discs. The center of the cartilage model of a developing vertebral body and its endochondral bone lie along a line extending (approximately) perpendicular to the central portion of the long axis of the closest section of the adjacent intervertebral disc. Hemivertebrae and bifid vertebrae form in relation to vertically oriented generally midline discs; the wedged vertebrae, and some hemivertebrae, form in relation to obliquely oriented discs; and bony bars (unilateral or block) form in relation to cartilage vertebral body models not separated by discs. The more abnormal the intervertebral discs, the more abnormal are the adjacent vertebral bodies. The final deformity is not fully patterned in the early embryo; it is due to a combination of primary genetic/epigenetic factors, secondary "obligate" mechanisms, and tertiary biomechanical forces.

Findings in pudgy mice involving all vertebrae and ribs and a case of severe human congenital scoliosis are remarkably similar. Vertebral morphogenesis is reviewed, including developmental concepts described over several decades and stressing current molecular concepts of variations in the oscillatory clock which patterns segmentation. An overview of human congenital scoliosis is outlined especially the severe deformities (spondylocostal dysplasias). The notochord is a precursor of the nucleus pulposus of the intervertebral disc, and the disc is invariably abnormal in relation to malformed vertebral bodies. Whether a convoluted notochord has a role in primary causation or is solely a passive displaced structure, its abnormal positioning could further induce vertebral body variability. Renewal of embryonic studies linking notochord/neural tube removal and transplantation experiments inducing vertebral malformation combined with assessments of gene/molecular components of the oscillatory segmentation clock and notochord should further clarify abnormal morphogenesis.

Keywords: Pudgy mouse; Histopathology; Vertebral, intervertebral disc, and rib abnormalities; Congenital scoliosis; Gene mutations; Segmentation clock; Notochord

Contents

1 Introduction 1

2 Materials and Methods 3
2.1 Source, Distribution, and Ages of Pudgy and Non-affected Mice 3
2.2 Whole Mount Preparations 3
2.3 Radiographic Studies 4
2.4 Histologic Studies 4
2.4.1 Serial Sections 5
2.5 Computerized Three-Dimensional Reconstructions 5
2.6 Previous Studies on Chick Vertebral Development and Human Congenital Scoliosis 5

3 Results 7
3.1 Gross Appearance 7
3.2 Whole Mount Appearances 7
3.2.1 Overview 7
3.3 Radiographic Studies 19
3.3.1 Overview 19
3.4 Histology Studies: Vertebrae, Ribs, Intervertebral Discs, and Ganglia 30
3.4.1 Overview 30
3.4.2 Serial Section Assessments 37
3.5 Computerized Three-Dimensional Reconstructions of Histologic Sections 56
3.6 Chick Embryo Vertebral Development (Previous Studies) 58
3.7 Radiology and Histopathology of Human Congenital Scoliosis of the Spine (With Normal Comparisons) 58
3.7.1 Normal Fetal and Neonatal Human Vertebral Bodies 58
3.7.2 Congenital Scoliosis of the Spine 60

4 Discussion 67
4.1 Normal Mouse Vertebral Development Compared to Pudgy Mouse Vertebral Development: General Considerations 67
4.2 The Study of Vertebral/Axial Development Showing Histologic and Molecular Progression 70
4.2.1 Concepts and Mechanisms of Embryonic Vertebral Development 70
4.2.2 Histogenesis of Vertebral Pattern Development 73
4.2.3 Timing in Embryonic Axial Development 77
4.2.4 Cell and Tissue Patterns in Vertebral Development: The Notochord and Other Regions Inducing Adjacent Embryonic Tissue Development 79
4.2.5 Rib, Ganglia, and Intervertebral Disc Development in Normal and Pudgy Mouse Specimens 80
4.2.6 Vertebral Development in the Chick Embryo 83
4.3 The Pudgy Mouse 84
4.3.1 Appearance 84
4.3.2 Gene Mutations 85
4.4 Genetic Influences on Axial Development: Mutations Identified in Mouse Models with Vertebral Deformation 86
4.4.1 Hox Genes 86
4.4.2 Mouse Vertebral Malformations 87
4.4.3 Mutations Affecting the Segmentation Clock and Wavefront: The Molecular Oscillator 88
4.4.4 Bovine Vertebral Malformations 88
4.5 Congenital Scoliosis (Human): Its Similarity with Pudgy Mouse Vertebral Abnormalities 89
4.5.1 Incidence and Associated Organ Abnormalities 89
4.5.2 Gross Anatomic, Radiologic, and Histopathologic Assessments in Human Congenital Scoliosis 89
4.5.3 Description and Categorization of Vertebral and Rib Abnormalities in Congenital Scoliosis 94
4.5.4 Correlation of Clinical, Radiographic, and Histopathologic Findings 95
4.5.5 Clinical Recognition of Congenital Scoliosis 95
4.5.6 Classification of Vertebral Disorders 96
4.5.7 Types (Patterns) of Vertebral Deformities 97
4.5.8 Abnormal Human Vertebral Development and Those Parts of the Spectrum Most Closely Aligned with the Pudgy Disorder 99
4.5.9 Genes Related to the Klippel-Feil Syndrome 99
4.6 Pathogenesis of Pudgy and Human Congenital Scoliosis Based on Histopathologic Studies 100

4.6.1 Vertebral Malformation: Mechanisms Underlying the Development of Structural Abnormalities in Congenital Scoliosis and Descriptive Classifications Based on Them 100
4.6.2 Role of Neural Tube/Spinal Cord and Notochord in Development of Axial Deformation: Vertebral Malformation Due to Errors of Induction 101
4.6.3 Relationship Between Shape and Positioning of the Intervertebral Discs (Nucleus Pulposus) [IVD/NP] and Adjacent Developing Vertebral Bodies in Pudgy Mice 104
4.6.4 Stages of Pathogenesis of Spinal Deformity 107
4.6.5 Summary of Rib Malformation 108
4.6.6 Summary of Intervertebral Disc Malformation 109
4.6.7 Summary of Ganglion Malformation 109

5 Conclusions 111
5.1 Implications of Pudgy Vertebral Abnormalities for Biologic Research 111
5.2 Implications of Pudgy Vertebral Abnormalities for Clinical Patient Treatment 112

References 115

Chapter 1
Introduction

Normal and abnormal vertebral development have been studied over the past 200 years at increasing levels of resolution as techniques for biological investigation have improved. Disordered development of the axial skeleton from the early embryonic period on leads to structurally malformed vertebrae and intervertebral discs and ribs causing the severe deformities of scoliosis, kyphosis, and kyphoscoliosis. Developmental malformation of the axial skeleton therefore has led to considerable biological and clinical interest. This work will detail our studies on the structural deformities of the vertebral column and adjacent ribs in the pudgy mouse [1] caused by mutations in the delta-like 3 (*Dll3*) gene of the Notch family [2]. While gene abnormalities in the pudgy mouse have been outlined, there has been no in-depth assessment of the histopathology of the pudgy vertebral and rib abnormalities that this study will provide. In addition, although congenital scoliosis has been recognized as a clinical problem since the mid-nineteenth century (1800s) [3] and accurately defined by radiography since the early twentieth century (1900s) [4–6], there have been few detailed histopathologic studies of human cases. We will also relate our histopathologic findings in the pudgy mouse to the histopathology of human vertebral and rib malformations in clinical cases of congenital scoliosis, one of which we defined in detail previously [7].

The pudgy mouse was recognized in the 1950s and described initially by Grünberg in 1961 [1]. The underlying recessive gene was induced by X-rays secondary to "specific locus" experiments carried out at the Oak Ridge National Laboratory in Tennessee (USA). The pudgy mouse was subsequently bred there and at Jackson Laboratories in Maine (USA). The entire pudgy mouse vertebral column is shortened with marked structural irregularities of the vertebrae, intervertebral discs, and adjacent ribs. The tail is shortened, twisted, and deformed due to the vertebral abnormalities. The bones of the appendicular skeleton are normal. A few affected newborns fail to thrive and die within two or three days of birth but the majority are systemically healthy, survive, and move actively about their cages.

In this study we concentrate on the structural changes of the vertebral column and ribs in pudgy mice compared with age-matched non-affected siblings. We

F. Shapiro, *Disordered Vertebral and Rib Morphology in Pudgy Mice*, Advances in Anatomy, Embryology and Cell Biology 221, DOI 10.1007/978-3-319-43151-2_1

outline the development of late embryonic and postnatal structural irregularities at varying ages. Multiple breeding pairs were used to accumulate the affected specimens. The pudgy mouse studies include whole mount preparations, high-resolution specimen radiographs, plastic-embedded toluidine blue-stained histologic sections, and computerized three-dimensional reconstructions of serial histologic sections.

The biological understanding of both normal and abnormal vertebral development over the past several decades has been furthered by investigational studies including (i) surgical interventions in the chick embryo (removal of specific regions/structures, grafts/transfers of tissue regions, and chick-quail chimeras), (ii) genetic mutants in the mouse both naturally occurring and transgenic (genetically modified), and (iii) clinical-radiographic studies in the human (with some histopathology and gene analysis). Normal and abnormal embryologic and early postnatal development in the chick, mouse, and human will be discussed in relation to pudgy mouse findings.

Chapter 2
Materials and Methods

2.1 Source, Distribution, and Ages of Pudgy and Non-affected Mice

Mice obtained for this study were products of pudgy breeding pairs from Jackson Laboratories, Bar Harbor, Maine. For the affected pudgy mice, heterozygous unaffected littermates served as controls. An affected pudgy mouse (pu/pu) can be identified at birth since it is approximately three-quarters the length of its non-affected littermates (pu/+) and has a markedly shortened, twisted tail. The mice were sacrificed by intraperitoneal injections of sodium pentobarbital. Vertebral and rib assessments were performed in 68 mice, 37 affected (pu/pu) and 31 non-affected (pu/+) age-matched siblings from the late embryo to 3 months of age. There were eight sets of births (litters) in which two or more of the sibling littermates were affected, allowing for a comparison of rib and vertebral anomalies in pudgy mice from the same mother and same pregnancy as well as with all other pudgy mice.

2.2 Whole Mount Preparations

Twelve mice underwent whole mount preparation, seven pu/pu and five pu/+. Following sacrifice the skin, heart, lungs, and abdominal contents were removed. Extremity muscles were not excised. The specimen was exposed to running cool tap water for 1 h, dried gently with paper towels, and placed in a fixative-stain solution composed of absolute alcohol (80 ml), acetic acid (20 ml), and alcian blue 8GX (Sigma Chemical Co., St. Louis, MO) (15 mg). The specimen was left in this solution for 24–48 h, the length of time being dependent on the intensity of alcian blue staining of cartilage portions of the skeleton. Following this the specimen was dehydrated in absolute alcohol for 5 days. Further staining was performed with

F. Shapiro, *Disordered Vertebral and Rib Morphology in Pudgy Mice*, Advances in Anatomy, Embryology and Cell Biology 221, DOI 10.1007/978-3-319-43151-2_2

alizarin red (Sigma) prepared using 10 mg alizarin red in 1 % potassium hydroxide. The staining continued until a deep red color indicated those skeletal parts that had converted to bone tissue. The specimen was then transferred for clearing to Mall solution composed of 79 ml water, 20 ml glycerin, and 1 g of potassium hydroxide. Once clearing of the adjacent soft tissues had been completed, the specimen was transferred to glycerin in increasing amounts from 70 to 80 to 90 %. The specimen was stored eventually in 100 % pure glycerol at room temperature in a closed drawer to minimize exposure to light that could lead to fading of the stain. Examination was performed to assess vertebral and rib patterns from dorsal, volar, and right and left lateral planes using a dissecting microscope (Carl Zeiss, Germany) equipped with a Contax RT5 electronic SLR system camera (Yashica, Japan). This method was adapted from those described previously [8–10].

2.3 Radiographic Studies

The vertebral column and ribs (primarily adjacent posterior and lateral ribs) were assessed using specimen radiographs. Radiographs were taken in 32 mice, 19 pu/pu and 13 pu/+. In those specimens that were to undergo whole mount preparation, the sternum and ventral portions of the rib cage were removed prior to radiography to allow clear visualization of the vertebrae and posterior ribs. In all other specimens, the vertebral column with the adjacent rib cage was excised in one piece from the base of the skull to the pelvis. Anteroposterior (ventral-dorsal) specimen radiographs were made using a General Electric Faxitron machine. Varying times and exposures were used for groups of specimens to obtain the sharpest images. The most common exposure used was for 48 s at 30 kV but times varied between 30 to 48 s and kilovolts used between 25 and 40.

2.4 Histologic Studies

Histologic studies were performed on 56 mice, 30 pu/pu and 26 pu/+. Following sacrifice and dissection, the spines and rib segments were fixed immediately in 10 % neutral buffered formalin. Fixation continued for 2 weeks following which the tissue segments were transferred to 25 % formic acid for decalcification. Once the tissues were soft, they were cut into smaller segments and infiltrated in JB4 medium (Polysciences, Warrington, PA) for several weeks. Tissues were embedded in JB4 plastic for sectioning. Sectioning was performed primarily in the coronal plane but additional sections were cut in the sagittal and transverse planes. Sections were cut at 5 μm and stained with 1 % toluidine blue.

2.4.1 Serial Sections

Serial sections (60–200) were made in 13 mice for qualitative study alone (10) and three-dimensional reconstruction (3), 11 in pudgy (pu/pu) mice and 2 in non-affected (pu/+) mice. Serial sectioning for qualitative study alone was done in the coronal plane in the spinal specimens from mid-thoracic to upper sacral vertebrae. The flattest vertebral region was the lower thoracic-lumbar part of the spine, minimizing the number of sections and making interpretation easier. Photomicrographs were taken at 3× magnification and printed at 21.6 × 27.9 cm.

2.5 Computerized Three-Dimensional Reconstructions

Three-dimensional computerized reconstructions were performed on serial histologic sections in one embryonic pu/+, one embryonic pu/pu, and one newborn pu/pu mouse. Serial sections were made in the sagittal plane of several segments of the thoracolumbar spine (vertebral bodies and intervertebral discs) encompassing the entire width of the column. Serial sections for reconstruction were photographed using a Zeiss photomicroscope at 10× magnification and printed to a uniform 12.7 × 17.8 cm. This allowed for outlining the differing anatomic components in development as the sections traversed the entire segment: the cartilage model of the vertebral body, bony part (ossification centers) of the vertebral body, nucleus pulposus of the intervertebral disc, and annulus fibrosus of the disc. A cursor traced the photomicrographs outlining tissue components. Reconstructions were performed on 60 photomicrographs/mouse vertebral column, evenly spaced from the entire width of the tissue being studied, using AUTOTRACE software. A color code was applied to each structure: the nucleus pulposus (yellow), annulus fibrosus (red), cartilage model of the vertebral body (dark blue), and vertebral body bone (light blue). A video camera sent images to a Gould frame rubber that was interfaced to a VAX 11/780. The images of ordered sections were then micro-aligned. The interactive computerized analysis reconstruction system was then used to display, rotate, and photograph the reconstructed images. These methods were adapted from the work of Harris and colleagues [11, 12].

2.6 Previous Studies on Chick Vertebral Development and Human Congenital Scoliosis

Selective results are reported from our previous studies on chick embryo vertebral development [13] and on findings in a human case of neonatal congenital scoliosis (with normal human fetal and neonatal spine comparisons) [7]. In the congenital scoliosis study, a continuous spinal segment from upper thoracic to lower sacral vertebrae was obtained. A specimen radiograph of the abnormal spine was taken prior to tissue preparation. The congenital scoliosis spinal segment was fixed in 10 % neutral buffered formalin and decalcified in 25 % formic acid prior to paraffin embedding, sectioning at 6 μm and staining with hematoxylin and eosin and safranin O-fast green. Seventy serial sections, cut in the coronal plane, were assessed. Intact normal vertebrae for histologic study had been obtained previously and processed the same way from an approximately 18-week-old fetus and from a 14-day-old neonate who died of encephalitis.

The study was approved by the institutional review board of Boston Children's Hospital.

Chapter 3
Results

Table 3.1 outlines ages of pudgy and non-affected mice, sibling groups with two or more littermates affected, and types of structural studies performed in each mouse. Table 3.2 outlines the age distribution of the studies.

3.1 Gross Appearance

An affected mouse is shorter than a non-affected one by approximately 25 % and, from birth onward, has a short curled tail in comparison with the longer straight tail of the normal. Photographs comparing normal (pu/+) mice and age-matched littermates of affected pudgy (pu/pu) mice are shown at late embryonic, newborn, and 6 weeks postnatal ages (Fig. 3.1a–c).

3.2 Whole Mount Appearances

3.2.1 Overview

Twelve mice underwent whole mount preparation; seven were affected mutants (pu/pu) and five were normal (pu/+). In the affected mutants, one specimen per time frame was assessed at birth (newborn), 8 days, 9 days, and 3 weeks, and three specimens were assessed at 4 weeks of age. In the normal mice, one specimen per time frame was assessed at birth (newborn), 8 days and 3 weeks, and two specimens were assessed at 4 weeks of age. In each pudgy (pu/pu) mouse, the pattern of vertebral and rib abnormalities was different from the other pudgy mice, including those affected siblings from the same mother (same litter, same pregnancy). Deformation was seen in virtually every cervical, thoracic, lumbar, sacral, and

F. Shapiro, *Disordered Vertebral and Rib Morphology in Pudgy Mice*, Advances in Anatomy, Embryology and Cell Biology 221, DOI 10.1007/978-3-319-43151-2_3

Table 3.1 Ages, distributions, and investigations in pudgy and non-affected mice

Sibling families	Mouse number	pu/pu	pu/+	Age	Xrays	Whole mounts	LM	Serial LM
	1	✔		3m	✔		✔	
↑	2	✔		2m	✔		✔	
	3		✔	2m	✔		✔	
S1	4	✔		2m	✔		✔	
	5	✔		2m	✔		✔	
↓	6		✔	2m			✔	
	7	✔		4h			✔	
	8		✔	4h			✔	
	9	✔		3d			✔	
↑	10	✔		3w	✔		✔	
S2	11		✔	3w	✔		✔	
	12		✔	6w	✔		✔	
↓	13	✔		6w	✔		✔	✔
	14		✔	9d	✔		✔	
	15	✔		9d	✔		✔	✔
	16		✔	1d			✔	
	17	✔		1d			✔	
↑	18		✔	nb	✔		✔	
	19	✔		nb	✔		✔	
S3	20	✔		1d	✔		✔	
	21		✔	1d	✔		✔	
	22		✔	1d			✔	
↓	23		✔	1d			✔	
	24	✔		nb	✔		✔	✔
	25		✔	nb	✔		✔	
↑	26		✔	3w	✔		✔	
	27	✔		3w	✔		✔	✔
S4	28		✔	3w	✔		✔	
	29	✔		3w	✔		✔	✔
	30	✔		3w		✔		
↓	31		✔	3w		✔		
	32	✔		8d		✔		
	33		✔	8d		✔		
↑	34	✔		9d		✔		
	35	✔		4w	✔		✔	✔
S5	36	✔		4w	✔		✔	
	37	✔		4w		✔		
	38		✔	4w	✔		✔	
↓	39		✔	4w		✔		

(continued)

Table 3.1 (continued)

Sibling families	Mouse number	pu/pu	pu/+	Age	Xrays	Whole mounts	LM	Serial LM
	40		✔	6w			✔	
	41	✔		4w			✔	
	42		✔	4w			✔	
	43	✔		6d			✔	
	44		✔	6d			✔	
	45	✔		4w	✔		✔	✔
	46		✔	4w	✔		✔	
	47	✔		nb			✔	✔
	48		✔	nb			✔	
↑	49		✔	3d	✔		✔	
S6	50	✔		3d	✔		✔	✔
↓	51	✔		3d	✔		✔	
↑	52	✔		nb	✔		✔	
S7	53		✔	nb	✔		✔	
	54	✔		nb	✔	✔		
↓	55		✔	nb		✔		
	56	✔		nb			✔	✔
	57		✔	nb			✔	
↑	58	✔		E			✔	
	59	✔		E			✔	
	60	✔		E			✔	
S8	61	✔		E			✔	✔
	62	✔		E			✔	
	63		✔	E			✔	✔
	64		✔	E			✔	
↓	65		✔	E			✔	
	66	✔		4w		✔		
	67	✔		4w		✔		
	68		✔	4w		✔		

Code: S = sibling; ↑ S ↓ = littermates (same mother) with 2 or more affected pudgy mice; pu/pu, pudgy mice; pu/+, non-affected mice; age = age at study; h = hours; d = days; w = weeks; m = months; LM = light microscopy study; Serial LM = serial histology sections.

tail vertebra in each pudgy mouse. Abnormalities involved hemivertebrae, wedged vertebrae, bifid (butterfly) vertebrae, and partially (unilateral bar) or completely (block) fused adjacent vertebrae. The rib cage was always abnormal bilaterally in asymmetric fashion. The intervertebral discs were irregular in shape and position; only rarely was the nucleus pulposus central and symmetric. There was often cartilage tissue continuity between adjacent vertebral bodies across the space where the intervertebral disc/nucleus pulposus would normally be positioned but was absent. Whole mount specimens are illustrated for normal (pu/+) mice in Fig. 3.2a–i and for pudgy (pu/pu) mice in Fig. 3.3a–f.

Table 3.2 Age distribution of studies

Age	pudgy mice (pu/pu)	non-affected mice (pu/+)		percentages
Youngest specimens (28 mice)				
Embryo to 1 day of age	**14**	**14**	=	**41%**
Entire series (68 mice)				
Embryo to 9 days	**21**	**18**	=	**57%**
3 weeks to 3 months	**16**	**13**	=	**43%**

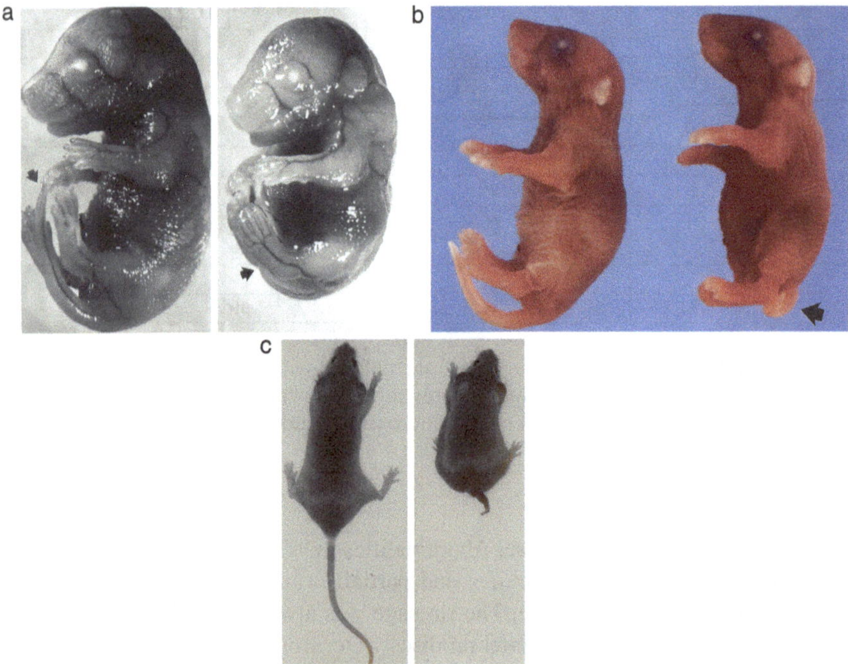

Fig. 3.1 Photographs show age-matched and litter-matched mice (pudgy, pu/pu, at the right and non-affected, pu/+, at the left) from: (**a**) late embryonic, (**b**) newborn, and (**c**) 6 weeks postnatal time periods. In the late embryo pu/pu mouse (**a**), the tail is short but not yet deformed (*curled*); in the newborn pu/pu (**b**), it is short and curled (*arrow*); and at 6 weeks (**c**) it is markedly shortened and deformed. The *arrows* point to the tails

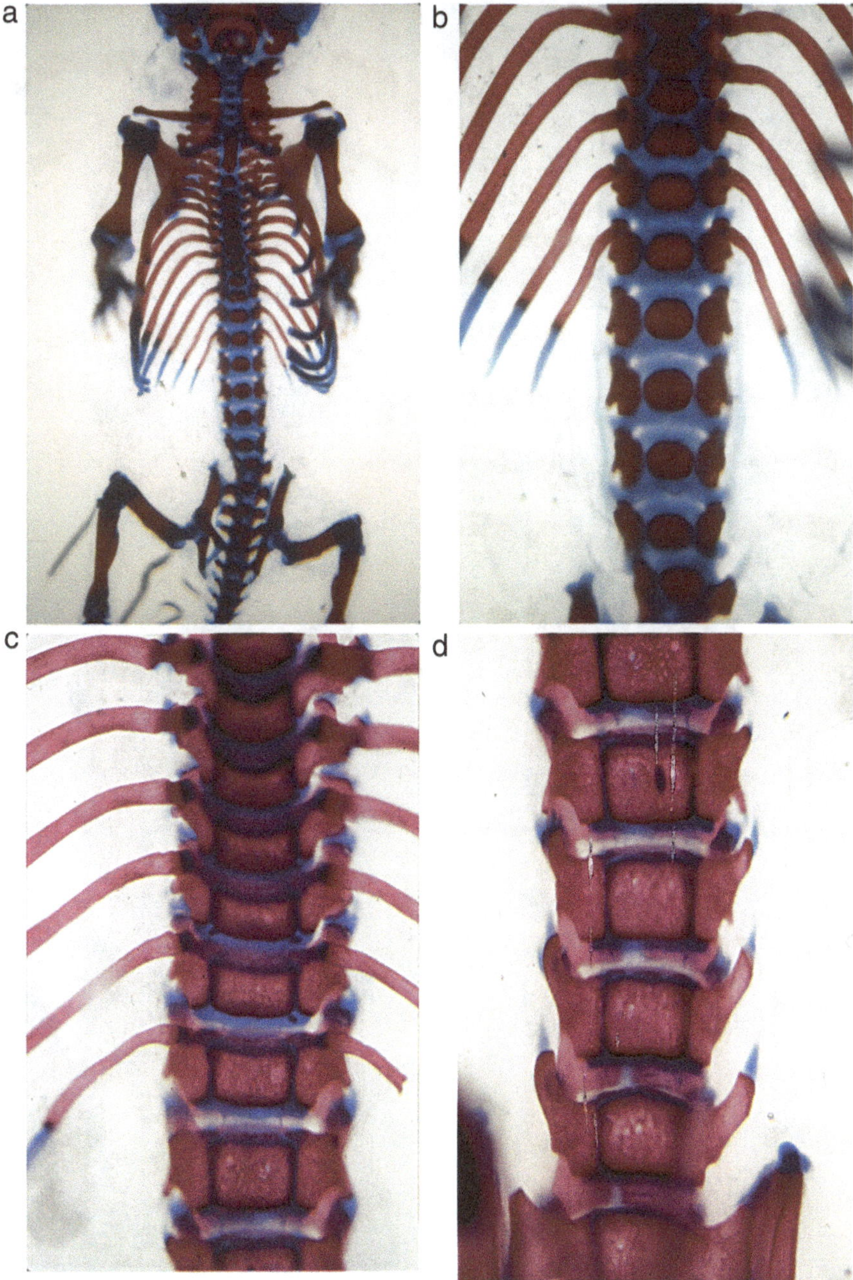

Fig. 3.2 Whole mount preparations of normal non-affected (pu/+) mice are shown. The cartilage tissue stains *dark blue*, the intervertebral disc tissue is either *light blue* or clear, and bone tissue stains *red*. The specimens in (**a**) and (**b**) are from newborn mice, while specimens in (**c**) to (**i**) are from 3-week-old mice. (**a**) The axial skeleton is seen in ventral-dorsal (anteroposterior) projection

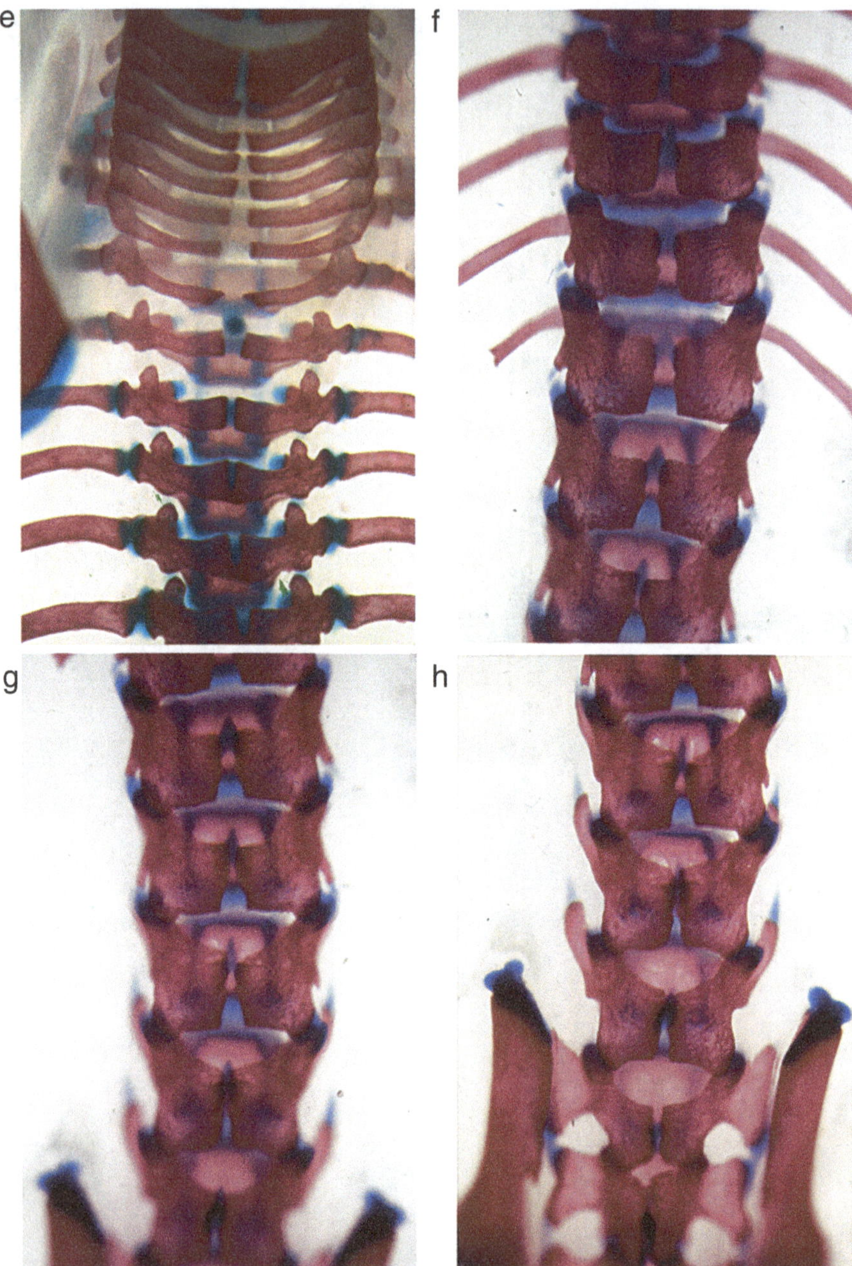

Fig. 3.2 (continued) from the cranio-cervical region at the top to the upper tail below the pelvis at the bottom. (**b**) The thoracolumbar region is seen with vertebral bodies showing central bone development in thoracic and lumbar vertebral bodies. Separation of ossification fronts (*red* staining tissue) is seen between the vertebral bodies (centrum) and the posterolateral regions with cartilage (*blue* staining tissue) intervening. (**c**) A segment of the thoracolumbar region is seen,

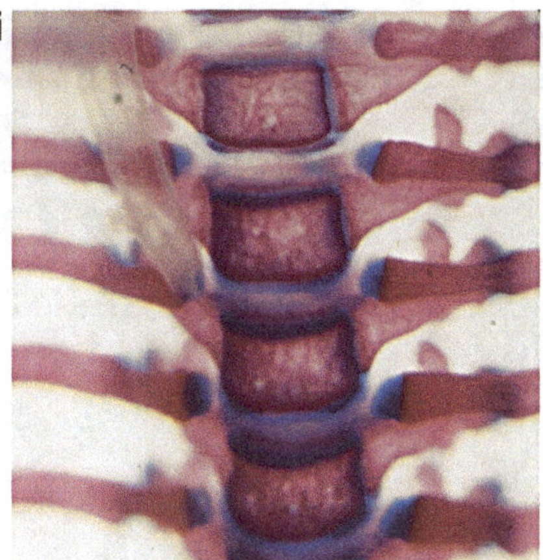

Fig. 3.2 (continued) photographed from the ventral-anterior surface of the vertebral bodies. The vertically oriented growth cartilage of the neurocentral physes is evident on both right and left sides. (**d**) The full lumbar segment of the spine including the upper region of the sacrum shows the same developmental features as in figure (**c**). The cartilaginous growth plates (*blue* staining) at the upper and lower vertebral body surfaces adjacent to the intervertebral discs are seen clearly. The pelvis is seen below it either side with the iliac bone (*red*) capped by the iliac apophyseal cartilage (*blue*). (**e**) A dorsal-posterior view focused on the neural arch of parts of the cervical and upper thoracic regions shows incomplete neural arch bone closure as well as vertebral body development. (**f**) A dorsal-posterior view of the thoracolumbar region also shows incomplete bony closure of the posterior neural arch. The ribs are symmetric. (**g**) Dorsal-posterior view of the lumbar region is shown. (**h**) Dorsal-posterior view of the lumbosacral region is shown. The iliac bones below stain *red* and their iliac cartilage apophyses stain *blue*. (**i**) Ventral-anterior view of the thoracic region focusing on costo-vertebral junctions shows normal symmetric relationship of developing ribs to the vertebrae. *Blue* staining cartilage of proximal rib regions relates to the vertebral bodies of two adjacent vertebrae on either side of the intervertebral discs. Each of the vertebrae in (**c**) to (**i**) has a cartilaginous (*blue* staining) vertically positioned neurocentral physis between the vertebral bodies (centrum) and the posterolateral neural arch bone bilaterally; a cartilaginous growth plate (*blue* staining) at the upper and lower vertebral body surfaces adjacent to and continuous with the intervertebral discs; and the cartilage (*blue* staining) of the facet joints seen most clearly in (**c**), (**d**), (**f**), (**g**), and (**h**)

3.2.1.1 Normal Development

Newborn

There are 13 pairs of symmetrical ribs. They are almost completely red staining throughout indicating the presence of bone tissue with only the ventral segments showing persisting blue cartilage. The vertebral bodies have prominent red staining central circular bone segments, but most of each vertebral body stains blue representing a preponderance of cartilage tissue. Even allowing for the slight relative size differences, the cervical vertebral bodies have proportionately less bone than those of the thoracic vertebrae. The whole mount preparation allows for clear differentiation between the nucleus pulposus and annulus fibrosus at each

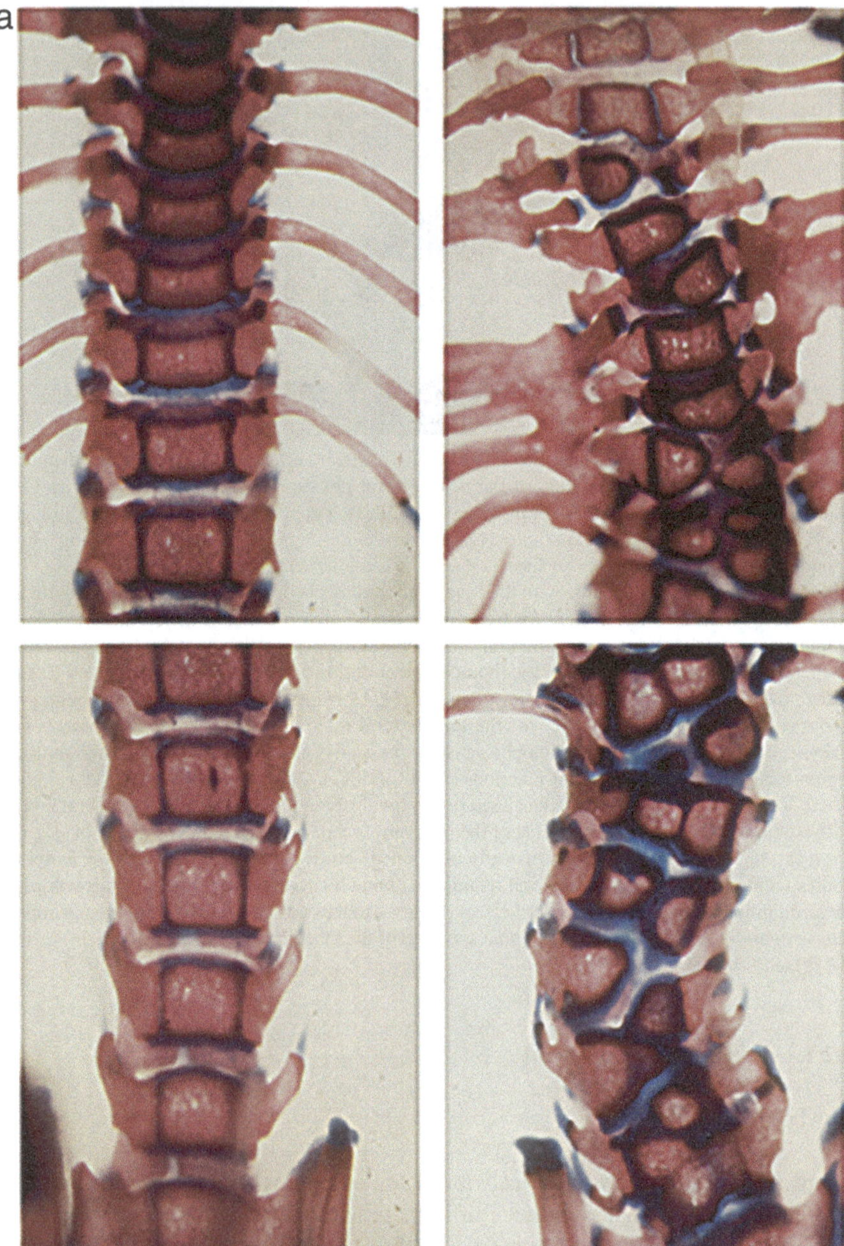

Fig. 3.3 Whole mount preparations of pudgy (pu/pu) mice are shown. (**a**) Whole mount preparations show the marked vertebral abnormalities from a 9-day-old pudgy (pu/pu) mouse (*right side panels*) compared with corresponding regions from a non-affected (pu/+) mouse (*left side panels*). Thoracic vertebrae are seen at the top and the lumbar below. On the right, blue staining cartilage tissue outlines the malformed pudgy hemivertebrae, wedged vertebrae, and bifid vertebrae; many

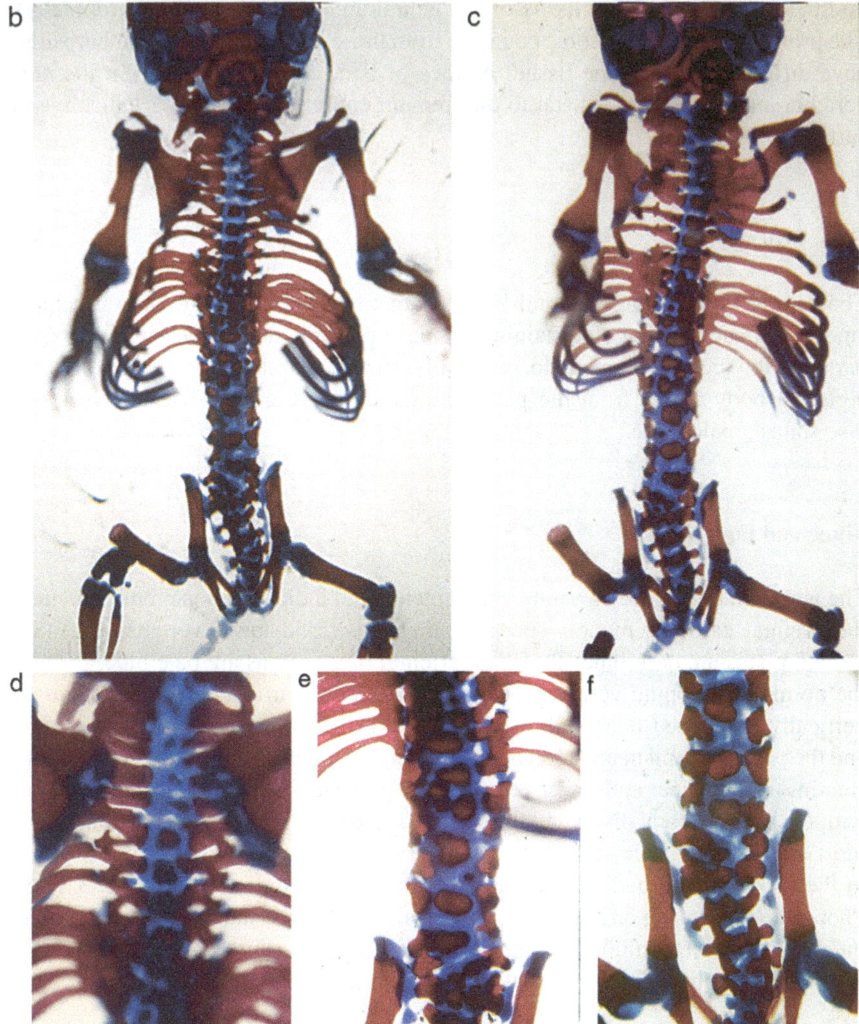

Fig. 3.3 (continued) intervertebral discs are obliquely oriented to variable degrees; and there are multiple rib abnormalities. (**b**) Whole mount preparation from a newborn pudgy (pu/pu) mouse demonstrates both vertebral and rib abnormalities. (**c**) Whole mount preparation from a newborn pudgy (pu/pu) mouse from the same litter as the mouse in (**b**) shows that the marked vertebral and rib abnormalities are structurally different from those that developed in the littermate. (**d**, **e**, **f**). These preparations show adjacent spinal regions from a 9-day-old pudgy mouse. (**d**) Abnormal thoracic vertebrae and ribs. (**e**) Abnormal lumbar vertebrae. (**f**) Abnormal lumbosacral vertebrae. The pelvis, hip joints, and proximal femurs are all normally shaped even though the adjacent lumbosacral vertebrae are markedly abnormal

intervertebral disc. The lighter regions of blue stain uptake represent the nucleus pulposus of the central portion of the intervertebral disc. The lateral and posterior segments of the vertebrae have clearly differentiated to bone tissue. There is no midline formation of bone tissue of the neural arches of the vertebrae (dorsally). The proximal five tail vertebrae counting from those distal to the ischial tuberosities have differentiated to bone tissue. A trace of bone formation is seen in the next vertebra and all vertebrae distal to that remain entirely blue staining indicative of cartilage tissue only.

Eight Days

The central bone of the vertebral bodies stains red, but now there are only small amounts of surrounding blue staining cartilage present on the superior and inferior parts of the body and also dorsolaterally (neurocentral physis) separating the vertebral body center from the pedicles of the neural arches. The intervertebral disc stains a paler blue.

Three and Four Weeks

The vertebral bodies are now almost completely red indicating bone replacement of the original cartilage model. The persisting upper and lower vertebral cartilage growth plates are dark blue. The intervertebral disc regions are pale blue. Each of the normal developing vertebrae in Fig. 3.2c–i has a cartilaginous (blue staining) vertically positioned neurocentral physis between the vertebral bodies (centrum) and the posterolateral neural arch bone bilaterally; a cartilaginous growth plate (blue staining) at the upper and lower vertebral body surfaces adjacent to and continuous with the intervertebral discs; and the cartilage (blue staining) of the facet joints seen most clearly in Fig. 3.2c, d–h. The proximal rib growth cartilage stains blue adjacent to the two vertebral bodies it relates to at the costo-vertebral joints (Fig. 3.2c, i). Photographs focusing on the posterior (dorsal) neural arch show bone development but also incomplete midline closure at cervico-thoracic (Fig. 3.2e), thoracolumbar (Fig. 3.2d), lumbar (Fig. 3.2g), and lumbosacral (Fig. 3.2h) levels.

3.2.1.2 Pudgy Mouse Development

Newborn

All vertebral bodies in the cervical, thoracic, lumbar, and sacral regions are abnormal in shape, although some show only minimal changes. Bone tissue development in the cervical vertebral bodies is markedly less than in the thoracic region. Many patterns of vertebral body abnormality are seen in random array involving hemivertebrae, wedged vertebrae, bifid vertebrae, and fused vertebrae. There is failure of fusion of the dorsal (posterior) neural arch midline elements throughout the spine. There are different levels of blue staining intensity differentiating cartilage areas of the vertebrae (dark blue) from the paler intervertebral disc regions. The intervertebral discs are irregular in shape. The nucleus pulposus is rarely central and symmetric. In many instances there is continuity between the cartilage of adjacent vertebral bodies across the space where normal intervertebral disc should be seen but instead is completely absent. All tail vertebrae distal to the ischium are formed only in cartilage. The tail vertebrae are markedly diminished in number and all are markedly deformed, accounting for the shortening and curling of the tail. The rib cage is abnormal bilaterally in asymmetric fashion. In one specimen the rib cage on the left shows a common paravertebral region of origin for six ribs with only the lower two having a normal conformation lateral to the point of paravertebral takeoff. The two ribs immediately proximal are fused focally adjacent to the vertebrae as are the two proximal to them. The next four proximal ribs are separate with normal-appearing points of origin although they are irregularly spaced. Asymmetric abnormalities are seen on the opposite right side. Certain areas where ribs should be present have none (described as rib deletion areas). The commonest general appearance is a linear paravertebral cartilage/bone mass from which as many as five separate ribs originate. This large paravertebral bone mass still attempts to relate to the adjacent vertebral bodies through cartilaginous (blue staining) costo-vertebral joints (Fig. 3.3b, c). Adjacent ribs are often fused at the point of takeoff but soon branch into two or three separate ribs more peripherally. Some of these ribs are twice as wide as normal. The long bones and pelvis are normal in appearance, including the hip joints composed of the acetabula and femoral heads.

Eight and Nine Days

Vertebral abnormalities are seen from the cervical region to the end of the tail. Vertebral bodies, stained in red and misshapen throughout, are surrounded by blue staining cartilage. Hemivertebrae, wedged vertebrae, bifid vertebrae, and fused vertebrae are seen. At the apex of the scoliosis deformation, on the concave side, no vertebral bodies are present, while hemivertebrae are present on the convex side. The nucleus pulposus of the intervertebral disc is clear with little dye uptake. There is failure of midline closure of the neural arch posteriorly in almost all vertebrae from the first cervical vertebra to the sacrum, but the open areas are diminished in width from those in the newborn. The posterior spina bifida is quite different from vertebra to vertebra in terms of shape and spacing between the two segments. Vertebrae in the tail are abnormally shaped but bone formation is occurring. The shortness and curliness of the tail are due to the individual vertebral abnormalities. In one segment, the long axis of the tail deviates from a straight line by 90°. The rib cage is abnormal with decreased numbers of ribs, fusions of ribs appearing to originate from abnormal paravertebral cartilage/bone accumulations, subsequent branching of fused ribs, irregular spacing, and a lack of right/left symmetry. The paravertebral bone accumulation generally shows attempts at normal costo-vertebral relations via cartilage articulations (Fig. 3.3e). Although the rib variations differ, both from side to side and from other abnormal embryos, the pattern involves a tendency to rib deletions and a common point of rib origin from a fused paravertebral mass with from two to six ribs extending from that mass. There are areas of deletions (absent ribs) in these regions as well. Some of the ribs appear normal; others are fused at their origin in paravertebral positions and are two to three times the normal diameter. Fused ribs often branch into two or three separate ribs that are relatively normal in shape. A frequent abnormality involves fused vertebrae with adjacent individual rib origin more lateral than normal or, in some instances, not occurring at all. The long bones, pelvis, and hip joints are normal in size and shape (Fig. 3.3f).

Three Weeks

Virtually all the vertebrae have been converted to bone tissue (red staining) with few blue cartilage areas defined except for the vertebral body growth plates. The intervertebral disc spaces are clear. Dorsal laminar abnormalities with failure of neural arch closure are seen. There are well-differentiated superior and inferior (cranial and caudal) vertebral body surface growth plates. In those vertebrae closest to normal, this growth plate is parallel to the upper and lower surfaces of the vertebral body. In more abnormally shaped vertebrae, the cartilage growth plates (blue) outline well the shape of the abnormal segments and the extent of the adjacent cartilage. In some segments the cartilage growth region is V-shaped terminating in the middle of the body instead of passing transversely across the entire width of the vertebra, providing an excellent marker of the cartilage model deformation. It is particularly marked where two adjacent vertebrae have fused partially with bone tissue (red) but not across the entire transverse diameter where the blue cartilage persists. The long bones and pelvis are normal.

Four Weeks

Vertebral and rib abnormalities are seen in each of three 4-week-old specimens. There is failure of dorsal arch closure of the lower thoracic and upper lumbar vertebrae but cervical and lower lumbar vertebral arches are closed dorsally, although the laminae are thinner than normal in the cervical region. Virtually all the cartilage tissue has been converted to bone tissue. Deep staining blue regions outline well the persisting growth cartilage and define the presence of bifid, hemi-, wedged, and fused vertebrae. In one specimen, complete failure of midline neural arch bone closure is seen although the bone segments are close. The long bones and pelvis are normal throughout.

3.3 Radiographic Studies

3.3.1 Overview

Specimen radiographs were taken of the vertebral columns and the adjacent rib cage in 32 mice, 19 affected (pu/pu) and 13 non-affected (pu/+) littermates, from newborn to 3 months of age. The radiographs were taken in the ventral-dorsal plane for each specimen [equivalent to anteroposterior plane in clinical terminology]. Individual pudgy mice were assessed along with multiple affected littermate siblings from the same mother in seven of the eight sibling groups (except for late embryonic mice). In each pudgy mouse, including affected same litter siblings, there are dissimilar bone abnormality patterns of vertebrae and ribs. *Vertebral shapes* include hemivertebrae, wedged vertebrae, bifid vertebrae (butterfly vertebrae or double hemivertebrae at the same level), and fused vertebrae. The *intervertebral discs* were noted to be vertical in position to the long axis of the vertebral column (between bifid and double hemivertebrae), oblique in position (between wedged vertebrae), either partially or completely absent (where two or more adjacent vertebrae fused into one bony continuum), and only minimally affected (when present between vertebrae which themselves were only minimally abnormal). The rib abnormalities include a decreased number of five or six single rib origins (proximal ribs) per side with subsequent double or triple branching to allow nine to ten ribs laterally (distally), asymmetric spacing of ribs with bilateral symmetry of the abnormal ribs never seen, abnormal proximal rib origins from paravertebral bone (originally the cartilage) accumulations, partially fused ribs along the body or shaft of the rib at vertebral and sternal ends, single ribs wider than normal, deletions (absences) of ribs, and diminished or increased spacing between ribs. Specimen radiographs from individual normal (pu/+) mice are illustrated in Fig. 3.4a–c, from individual pudgy (pu/pu) mice in Fig. 3.5a–h and from comparative age- and litter-matched normal and pudgy mice in Fig. 3.6a–g. Examples of pudgy tail vertebrae are shown in Fig. 3.6h and rib/vertebral relationships in Fig. 3.6i–l.

3.3.1.1 Normal Development (Non-affected pu/+ Mice)

Non-affected littermates show the normal pattern of rib and vertebral development with 13 pairs of ribs, 7 cervical vertebrae, 13 thoracic vertebrae, 6 lumbar vertebrae, 4 sacral vertebrae, and numerous tail vertebrae (30–31).

Newborn [#s 18, 25, 53]

The most extensively ossified vertebral bodies are in the lower thoracic and upper lumbar regions with the lower lumbar and cervical regions proportionately less ossified. There is no ossification in any of the superior or inferior vertebral end

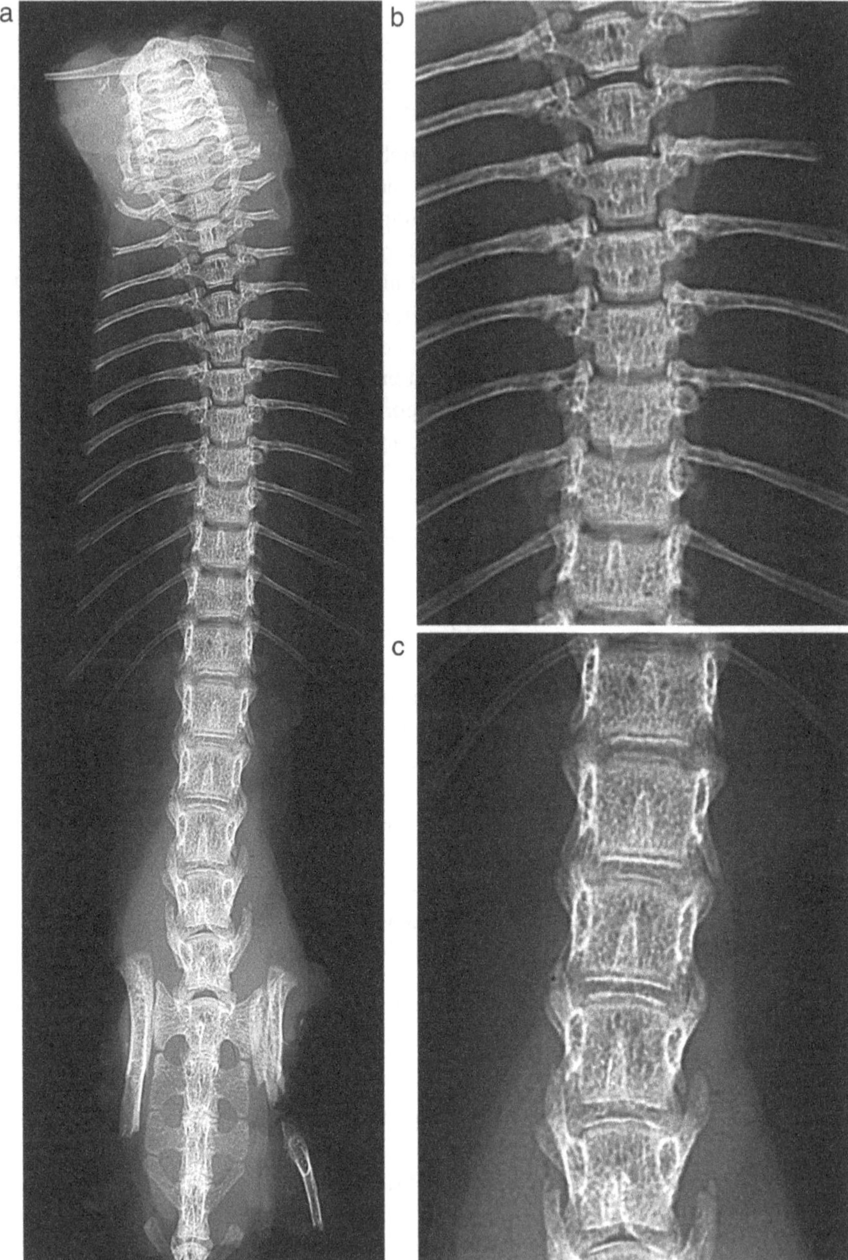

Fig. 3.4 Radiographic views of a non-affected (pu/+) 4-week-old mouse vertebral column with adjacent ribs in a ventral-dorsal (anterior–posterior) projection are seen: (**a**) entire spine, (**b**) thoracic region, and (**c**) lumbar region. Upper spinal curvature is due to the positioning of the specimen and slightly oblique projection rather than scoliosis. Note the thin linear radiodensity at the upper and lower surfaces of the upper thoracic and all lumbar vertebral bodies in (**b**) and (**c**).

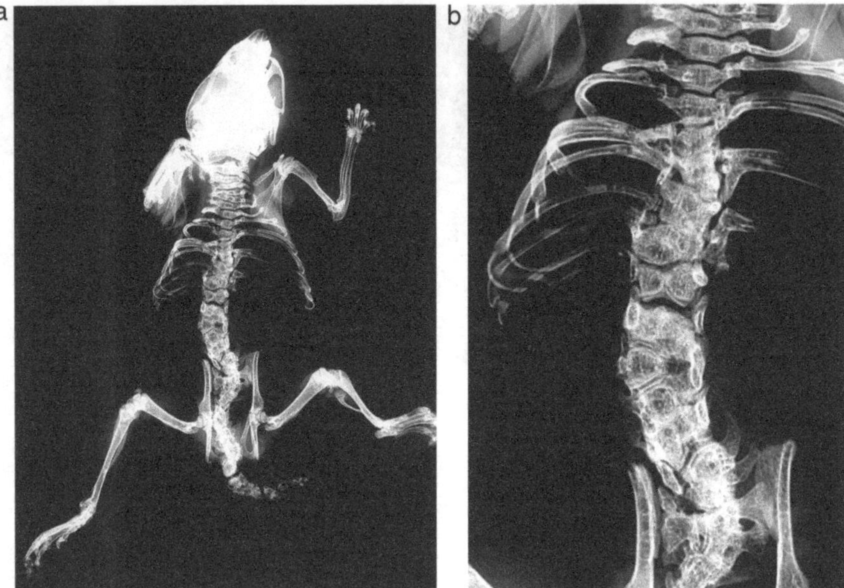

Fig. 3.5 Vertebral radiographs with adjacent ribs from several pudgy (pu/pu) mice are shown: (**a**), entire pudgy mouse skeleton at 4 weeks of age is visualized. Each vertebra from the cervical region to the tip of the tail is malformed. The pelvis and hip joints are normal, as is the rest of the appendicular skeleton. (**b**), vertebral column and ribs alone are shown from the same mouse in (**a**). Note the presence of the vertebral end plate thin radio-dense lines which serve to outline the limited growth extent of the malformed vertebrae; (**c**), 3 weeks of age; (**d**), 9 days of age; (**e**), 4 weeks of age; (**f**), 6 weeks of age; (**g**), 2 months of age; and (**h**), 3 months of age. The vertebral images show markedly different patterns of vertebral and rib malformation, even though the mutation is presumed to be the same in each. The paravertebral bone mass from which the asymmetric, fused ribs pass outward is often separate from the vertebral body walls in younger pudgy mice (**b**, **e**), but in older mice the mass appears continuous with the malformed vertebral bodies as in (**f**) (lower right but space seen on lower left), (**g**) (*right side*), and (**h**) (*right side*)

plates. The thoracic vertebral bodies possess an oval bone center. Cartilage spaces can be seen superior, inferior, and lateral to the ossified vertebral body throughout the thoracic and lumbar regions. Lumbar vertebral body region bone shape shows a transition from oval at the upper regions to circular at the lower. There are 13 pairs of ribs, symmetrical in appearance.

Fig. 3.4 (continued) This represents calcification and not secondary center ossification within the growth cartilage adjacent to the intervertebral disc, since no bone is seen within the cartilage on histologic sections. These linear densities are present in all thoracic vertebrae but are not seen in some radiographs due to the line of projection. The ribs are symmetric and evenly spaced on right and left sides. Note normal relationship of each rib to two adjacent vertebrae on either side of the intervertebral discs

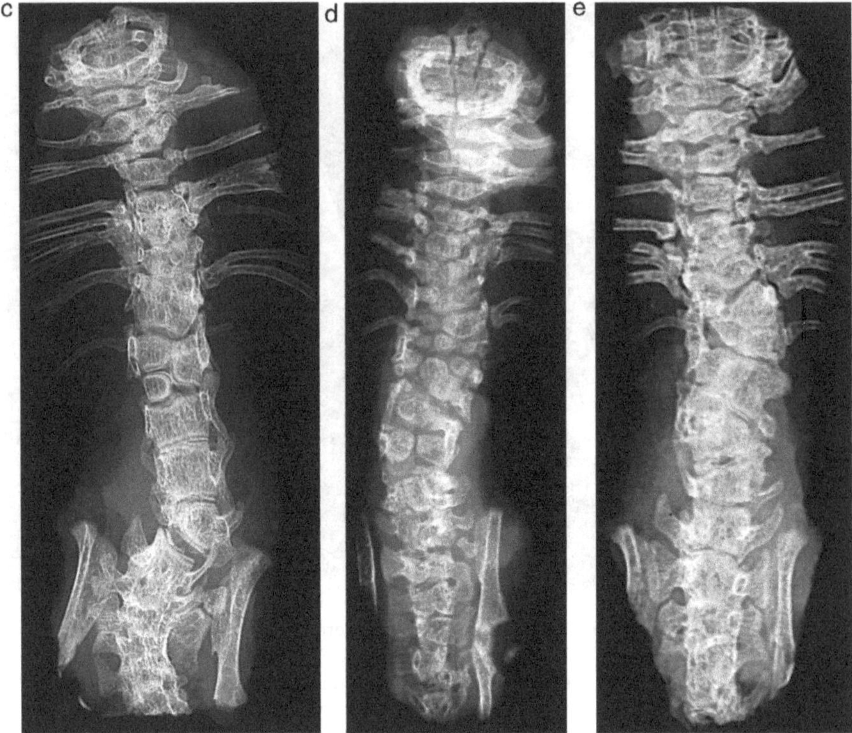

Fig. 3.5 (continued)

One Day [#21]

The ossification centers in the fourth and fifth cervical vertebrae have appeared but are still quite small. The findings are otherwise the same as in the newborn.

Three Days [#49]

There is no linear density present within the superior (cranial) or inferior (caudal) vertebral body growth cartilage regions.

Nine Days [#14]

Early traces of linear radiodensity are seen in the superior and inferior vertebral growth cartilage regions in thoracic and lumbar areas (positioned between the radiolucent intervertebral disc and the radiolucent growth cartilage). The right and left vertical vertebral body radiolucencies (neurocentral physes) are clearly visualized in the lumbar vertebrae but are not seen in the thoracic.

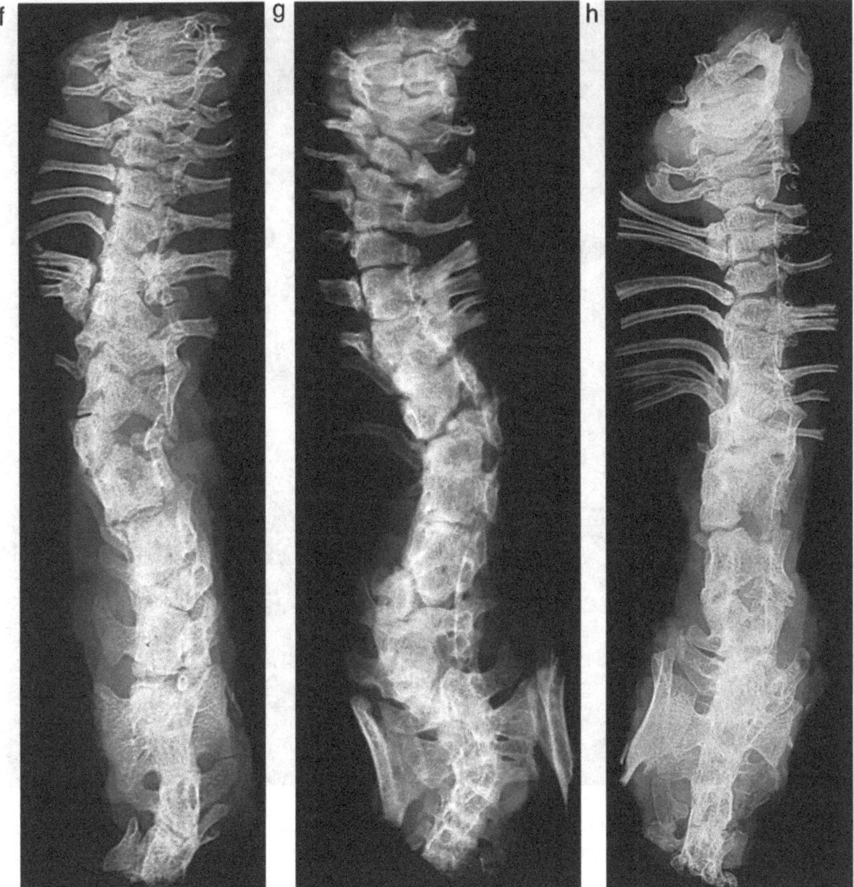

Fig. 3.5 (continued)

Three Weeks [#s 11, 26, 28]

Radiolucent cartilage growth plates at the inferior and superior surfaces of the vertebral bodies continue to be identified throughout the thoracic and lumbar regions, as do the linear radiodensities.

Four Weeks [#38]

Superior and inferior vertebral body growth plate radiolucency is noted throughout the thoracic and lumbar regions (Fig. 3.4a–c) along with the linear radiodensities described above. These are seen most clearly (due to the plane of radiation) in the upper thoracic (Fig. 3.4b) and all lumbar (Fig. 3.4c) vertebrae.

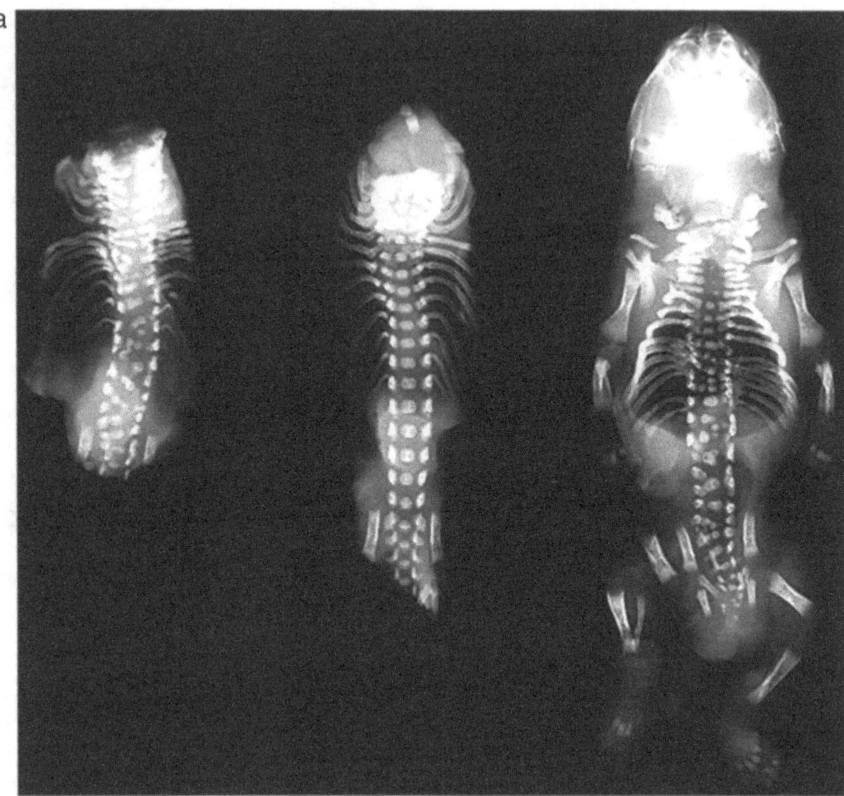

Fig. 3.6 Multiple specimen radiographs of non-affected and pudgy mice vertebral columns, many with age-matched spinal images of siblings, are shown in (**a**) to (**h**). Examples of rib-vertebral relationships in pudgy variants are seen in (**i–l**). (**a**) Radiograph illustrates three newborn mice from the same litter showing the vertebral columns, from left, of pudgy, non-affected, and pudgy (whole body) mice, respectively. (**b**) Radiograph illustrates two newborn mice from the same litter. (**c**) Radiograph illustrates two newborn mice from the same litter. (**d**) Radiograph illustrates two 1-day-old mice from the same litter. (**e**) Radiograph illustrates three 3-day-old mice from the same litter showing the vertebral columns, from left, of non-affected, pudgy, and pudgy mice, respectively. (**f**) Radiograph illustrates two 3-week-old mice from the same litter. (**g**) Radiograph illustrates two 3-week-old mice from the same litter. (**h**) Radiograph illustrates pudgy (pu/pu) spine. Also shown is the pelvis and tail from the same specimen. Note marked vertebral abnormalities in the tail that account for the marked shortening and deformation. The pelvic bones, hips, and proximal femurs are normal in shape. Figures (**i–l**) all show focal views of abnormalities of the ribs in pudgy mice. In each image throughout Figures (**a–l**), examination of most double or triple rib origins from a single paravertebral bone mass shows clear examples of a space between the paravertebral bone mass and the adjacent vertebrae consistent with efforts at formation of a joint. A few regions appear to show bone continuity between this paravertebral bone mass and the vertebrae, but this might be due to having the pseudarthrosis obscured by rotation at the site giving a picture of bone continuity. The same interpretation applies to the pudgy radiographs in Fig. 3.5

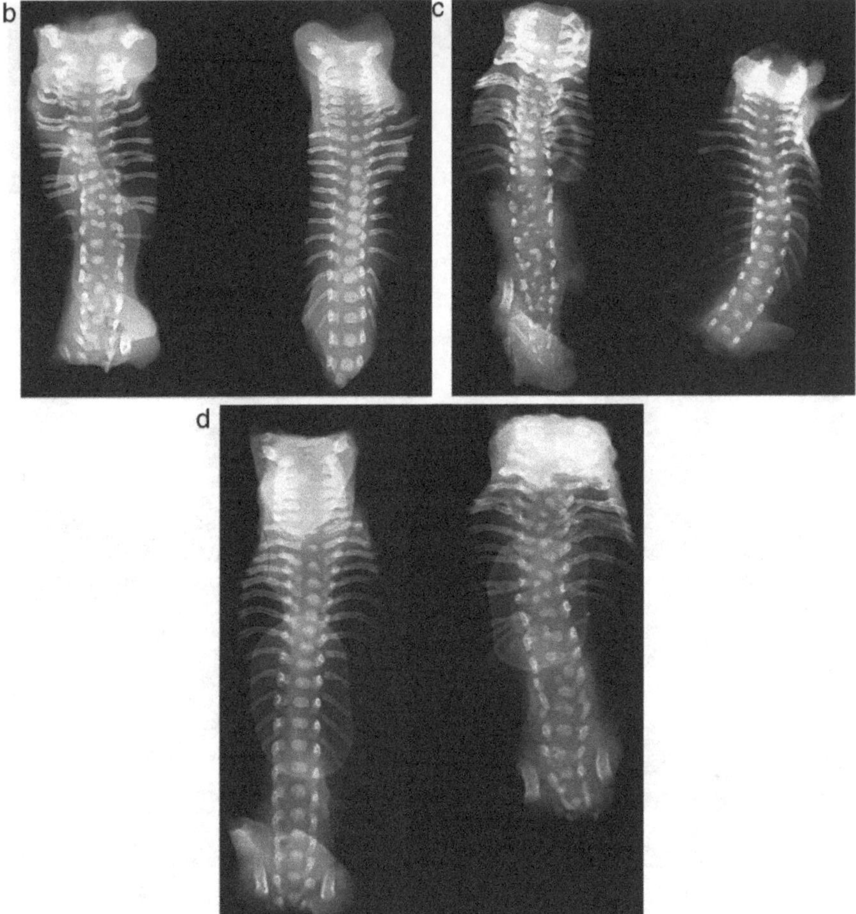

Fig. 3.6 (continued)

Six Weeks [#12]

The six lumbar vertebrae in this specimen are clearly visualized. Superior and inferior vertebral body growth regions are becoming thin and barely visible, indicating growth function slowdown.

Two Months [#3]

Further development of bone maturity is seen.

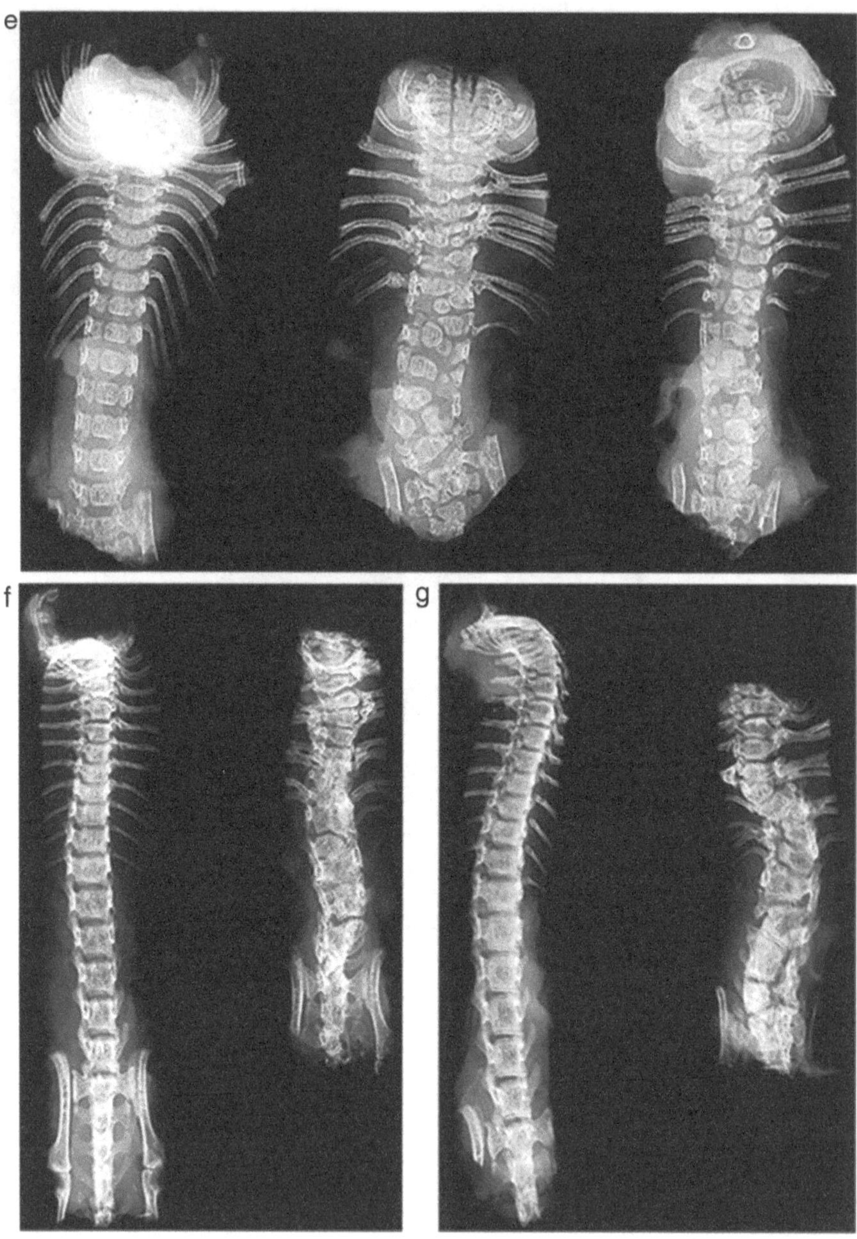

Fig. 3.6 (continued)

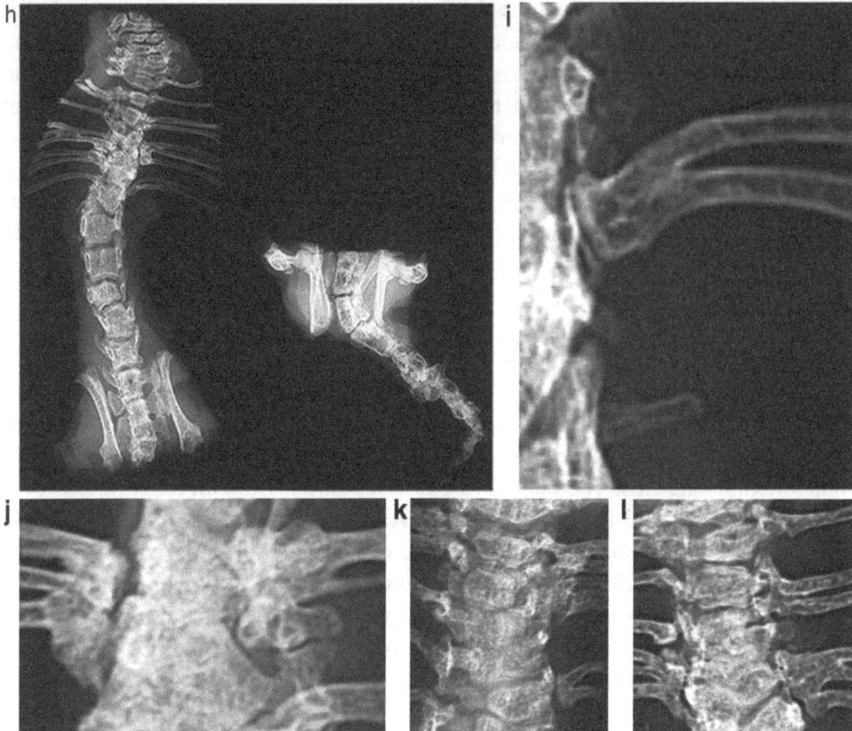

Fig. 3.6 (continued)

3.3.1.2 Pudgy Mouse Development (pu/pu Mice)

Overview

Abnormalities of ribs and vertebrae are seen in each pu/pu mouse. The specific patterns of irregularities however are different in each mouse assessed, including those siblings born from the same mother. Vertebral abnormalities are present throughout each region—cervical, thoracic, lumbar, sacral, and tail. Rib fusions and absences (deletions) are seen bilaterally. There are multiple fused areas of the spine (bone to bone continuity without cartilage or disc tissue interposition): (i) between relatively normal shaped adjacent vertebral bodies creating block vertebrae (completely uniting two vertebrae with no intervening intervertebral disc or cartilage tissue) or unilateral bony bars (uniting two vertebrae with some cartilage/disc tissue persisting) or (ii) between adjacent hemivertebrae or wedged vertebrae. The overall vertebral column alignment in one pudgy mouse can be straight showing no scoliosis in the anteroposterior (dorso-ventral) plane, while another sibling from the same litter has vertebral abnormalities causing a severe scoliosis deformation. The number of ribs is decreased from the normal, for example, in one mouse on one side, there are nine individual ribs seen with some branching from a single paravertebral origin of continuous cartilage tissue transformed to bone tissue with time. In most instances, and especially in the younger mice, there is a clear vertically oriented radiolucent space between this paravertebral cartilage/bone tissue mass from which the multiple ribs pass outward (Figs. 3.5b, e and 3.6i, j, left side, k, lower right side, l both lower sides). In some of the older mice, however, there appears to be bony continuity between the lateral vertebral bodies, the bony paravertebral mass, and the malformed ribs (Fig. 3.5f–h).

Findings in Vertebral/Rib Patterns Within Sibling Groups

In each of the pudgy mice within the seven groups of affected siblings where radiographs were taken, there are dissimilar bone abnormality patterns involving both vertebrae and ribs. The patterns are also dissimilar from all other pudgy mice assessed. Instead of 13 individual paired ribs, there are generally five to six single rib origins with subsequent double, triple, or quadruple branching (coalitions) with nine to ten individual ribs eventually present more peripherally (laterally, distally). Bilateral symmetry in the abnormal ribs is never seen. The patterns differ from mouse to mouse, including between siblings of the same litter. The rib deletions are irregular in number, spacing, and position, and the fusions (coalitions) are also variable.

Sibling Group 1 [#s 2, 4, and 5, Two Months]

In the Pu-2 (pu/pu) spine, there is a particularly marked double scoliosis curve noted secondary to the multiple bone abnormalities, while the vertebral columns of the other two mice are relatively straight.

Sibling Group 2 [#10, Three Weeks and #13, Six Weeks] #10

There are three relatively normal mid-thoracic vertebrae. The lower thoracic vertebrae over four segments are fused into a solid mass. The lumbar region shows a hemivertebra, a bifid vertebra, and a fusion of several lower vertebrae that are also misshapen. Even allowing for the three additional weeks of growth in specimen #13 compared to specimen #10, the exact pattern in particular in relation to the ribs is not reproduced between the two mice. The vertebral abnormalities in particular at the thoracolumbar junction also show different patterns in individual mice.

Sibling Group 3 [#19, Newborn and #20, One Day]

There are clear differences between the two mice with no similarity in shape and position of ossified vertebral bodies or ribs. The ribs in each mouse are markedly different. In #19 on one side, there are eight rib origins including two coalitions each of which branches to two ribs leading to a total of ten ribs more laterally. On the opposite side, there are six origins of which the three show multiple ribs leading to a total of ten ribs. In mouse #20 on one side, there are 12 ribs with separate but fused origins composed of three, two, two, and two ribs instead of normal separate individual origins. On the opposite side, there is a double origin and one involving three or four ribs that originate from one large fused paravertebral bone mass. Vertebrae throughout the thoracic and lumbar regions are abnormally shaped.

Sibling Group 4 [#s 27 and 29, Three Weeks]

The lowest ribs at the thoracolumbar junction appear different in #27 and #29. There are marked fusions into one irregular mass spanning the lower thoracic and upper lumbar regions. In #27, there are irregular intervertebral spaces, while in this same region in #29, there is a large unilateral bar and a wedged vertebra. There is a sharply angled scoliosis in the mid-thoracic region in #29 but an almost straight spine in #27. In the reasonably normal-appearing vertebrae in the pudgy variants, the superior and inferior growth plates are present with linear regions of radiodensity within each plate indicative of the calcified cartilage.

Sibling Group 5 [#s 35 and 36, Four Weeks]

There is a clear difference in the radiographs involving both vertebrae and ribs. The most evident is a sharply angled scoliosis curvature at the lower lumbar and sacral region in #35, while the alignment in lumbar and sacral vertebrae in #36 is straight. There is a hemivertebra at the apex of the curve at the lower lumbar region in #35 with no such finding in #36. The abnormal rib patterns are different both on right and left sides of each individual and in comparison between siblings #35 and #36. The radiographs demonstrate the cartilage growth plates at the superior and inferior vertebral body surfaces. On occasion these are intact across the entire diameter of the vertebral body. In other instances the cartilage plates stop abruptly leading to the impression of an associated trans-intervertebral disc bone bridge. There tends to be a solid mass of fused vertebrae at the thoracolumbar junction—even though one can make out irregular modeling distinct from three or four block vertebrae of regular margins.

Sibling Group 6 [#s 50 and 51, Three Days]

The patterns of rib coalitions and deletions are different both between the right and left sides of each rib cage and between the two specimens. There is an abrupt lower lumbar and sacral scoliosis in #50 but a reasonably straight alignment of vertebrae in #51. The combinations of vertebral shape changes at the thoracolumbar junction in each specimen are also markedly different. This is one of the clearest examples of the differing patterns in the study (Fig. 3.6e, middle and right-sided images).

Sibling Group 7 [#s 52 and 54, Newborn]

The rib patterns are markedly different. There is a double scoliosis (one thoracic curve convex to the right and a lumbosacral curve convex to the left) with multiple vertebral abnormalities. Marked abnormalities of individual vertebral shape are present throughout the thoracic and lumbar regions. No normal vertebrae are seen. The abnormalities include hemivertebrae, wedged vertebrae, bifid vertebrae, unilateral bars, and block vertebrae (Fig. 3.6a, pudgy siblings at the left and right with non-affected sibling in the middle).

3.4 Histology Studies: Vertebrae, Ribs, Intervertebral Discs, and Ganglia

3.4.1 Overview

Histology studies were performed on 56 mice, 30 affected (pu/pu) and 26 - non-affected (pu/+). Findings are illustrated in Figs. 3.7, 3.8, 3.9, 3.10, and 3.11. Figure 3.7a–f5 shows the developmental sequence of the vertebrae and intervertebral discs in normal mice; Fig. 3.8a–i, the abnormal vertebrae and intervertebral discs in the pudgy mice; Fig. 3.9a1–a5, normal rib development, and Fig. 3.9b1–b13, abnormal rib development in pudgy mice; Fig. 3.10a1–a6, normal ganglia development, and Fig. 3.10b1–b4, abnormal ganglia development in pudgy mice; and Fig. 3.11a1–a6, normal intervertebral disc development, and Fig. 3.11b1–b3, abnormal intervertebral disc development in pudgy mice.

Development of abnormal cartilage (and bone) models in the pudgy (pu/pu) mice reveals that bone tissue develops by normal endochondral and intramembranous mechanisms in malformed vertebrae, but bone tissue (considered as an organ) demonstrates abnormal shaping and position because of the abnormal cartilage models on which it has been synthesized. That being said, the bone always develops initially within the abnormally shaped cartilage models via 360° centers of endochondral ossification and only forms by intramembranous ossification when the bone tissue reaches the most peripheral extent of the cartilage model of any particular vertebra. The characteristic misshapen cartilage models lead to hemivertebrae, wedged vertebrae, bifid vertebrae (butterfly vertebrae or double hemivertebrae at the same level), and fused vertebrae. *Development of intervertebral discs* shows the same variability phenomenon, being abnormal in the sense of (i) passing only partially across the entire width between two developing vertebral bodies, (ii) not being present at all in what should be the intervertebral space between adjacent misshapen vertebral bodies, (iii) being obliquely or vertically positioned instead of transverse to the long axis of the vertebral column, (iv) being thicker or thinner than normal, or (v) being misshapen into round, hexagonal, curvilinear, or other patterns. *Development*

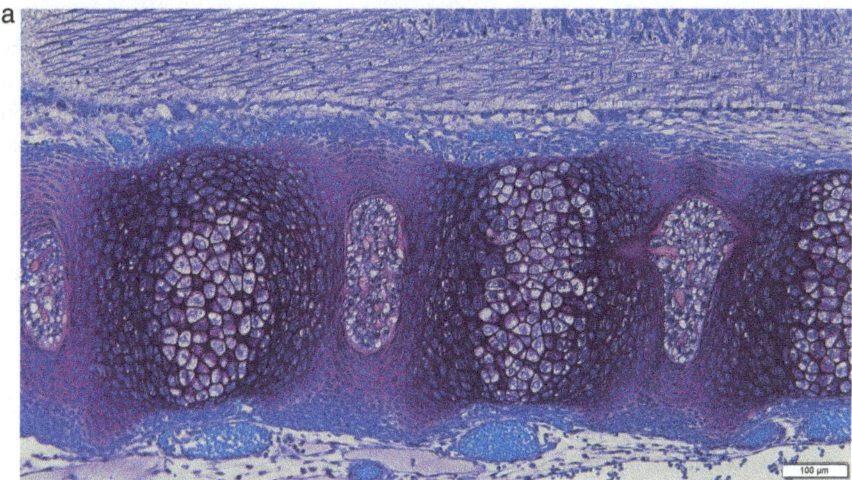

Fig. 3.7 Photomicrographs of normal (pu/+) vertebrae and intervertebral discs are illustrated. All sections were embedded in JB4 plastic and stained with 1 % toluidine blue. (**a**). Embryo. Vertebrae from thoracolumbar region show regular segmented vertebra—intervertebral disc formation with the spinal cord at the top. At the right, the intervertebral disc shows a receding notochordal fibrous remnant adjacent to the vertebral body. The vertebral bodies show central hypertrophic chondrocytes as the endochondral bone formation system proceeds but no actual bone has yet been synthesized in this section [This image is placed in the human horizontal position with cranial-head region to left and caudal region to right.] Marker line = 100 µm. (**b**1). Newborn. Photomicrograph of the lumbar vertebral body cut in the mid-coronal plane is shown with intervertebral disc above and partial disc below. The growth plates at the upper and lower surfaces of the body merge with the fibrocartilage of the developing intervertebral discs. Within the vertebral bodies, bone tissue is deposited on the cartilage cores via endochondral bone synthesis. At the right and left sides, the walls of the vertebral body are forming with the intramembranous bone. The *darker purple* staining regions of the growth plates and cartilage within the marrow represent the calcified cartilage. There is no bone formation between the disc and the cartilage of the growth plate. Marker line = 100 µm. (**b**2) Thoracic vertebral bodies with ribs bilaterally are shown. The plane of sectioning is from just posterior to the posterior vertebral body wall at the top of the image (into the tissue surrounding the spinal cord) and then passing at progressively lower levels through the posterior wall of the body in the next vertebra (intramembranous bone), completely within the vertebral body (endochondral bone), and, at the bottom, through the cartilage of a partial vertebral body. This plane of sectioning affects the appearance of the shafts of the ribs at the right side (the intramembranous bone of the rib cortex at the top, mixture of the intramembranous bone and part of the metaphyseal endochondral bone in the *middle*, and mid-rib section below with continuous marrow from the metaphyseal region to the far right of the image). Marker line = 200 µm. (**c**1). Newborn. Vertebral bodies, intervertebral discs, and ribs in the thoracic region are seen. Ossification of the vertebral bodies has started with endochondral bone formation in a 360° orientation. The annulus fibrosus at the periphery of each disc is becoming oriented. The ribs are beginning to articulate with the vertebral body above and below. Synovial joint formation is beginning at some rib-vertebral sites with cell resorption leaving a clear area, but most regions still show continuous interzone cellularity. Marker length = 200 µm. (**c**2 and **c**3). Newborn. Higher-power views of costo-vertebral joints show some regions of the joints with continuous cellularity across the interzone region, while some clear zones represent early cavitation due to interzone resorption. Marker lengths = 100 µm. (**d**1). 6 days. Photomicrograph within the lumbar region, in the mid-coronal plane, centered on an intervertebral disc shows the vertebral

of ribs in the pudgy mice was always abnormal. Major findings included decreased numbers of ribs, asymmetric spacing between ribs, deletions of ribs with no formation leading to wide spaces between adjacent ribs, fusion of two or more adjacent ribs with a continuous widened physis forming misshapen structures, subsequent branching of these initially fused ribs into separate ribs, apparent origin of these fused ribs from separate paravertebral cartilage accumulations that often extended the length of three or four adjacent vertebral bodies, and a tendency for these paravertebral cartilage collections to form costo-vertebral synovial-type joints via the joint interzone cellular transformation mechanism. *Development of ganglia* appeared to have normal cell components, although the shape and position were abnormal in the affected mice. Areas of ganglion continuity were seen in the pudgy mice with the ganglia misshapen and not aligned in normal longitudinal array or with regular spacing. Examination at higher levels of magnification always reveals a thin fibrous septum between each ganglion.

Fig. 3.7 (continued) body above and below. The nucleus pulposus of the disc contains remnants of notochordal cells. The annulus fibrosus of the intervertebral disc is forming well. The sidewalls of the vertebral bodies are forming with the intramembranous bone. Marker length = 100 μm. (**d**2). 6 days. Photomicrograph is centered on the vertebral body that was seen at the top of the previous image (**d**1). Marker length = 50 μm. (**d**3) A rib forming at a costo-vertebral junction in a newborn is shown. The synovial cavity has not yet formed in this section. Marker length = 100 μm. (**e**) Photomicrograph from a 9-day-old pudgy mouse is shown**.** A portion of a lumbar vertebral body in its posterior aspect is seen following sectioning of bone tissue in the coronal plane. The neurocentral epiphyseal cartilage plates in the vertical plane separate the vertebral body from the posterolateral pedicle bone at either side. These allow for bidirectional growth expansion on both body and pedicle sides. Marker length = 100 μm. (**f**) Images from 3-week-old pudgy mice vertebral bodies and intervertebral discs are shown. (**f**1) Photomicrograph centered on a vertebral body shows partial intervertebral discs above and below. The growth plates at the upper and lower surfaces of the vertebral body persist, but almost all the calcified cartilage in the vertebral body marrow has been resorbed with trabeculae now composed of the bone tissue only. Marker length = 100 μm. (**f**2). A well-formed intervertebral disc is shown from another mouse at the same age. Plane of sectioning intersects part of the posterior wall of the vertebral body (the intramembranous bone at the top). The cellularity of the nucleus pulposus is markedly diminished. The fibrous lamellae of the peripheral annulus fibrosus are well formed. Marker length = 200 μm. (**f**3) Vertebral body and disc are shown following sectioning partially within the body wall where the intramembranous bone is seen. Marker length = 100 μm. (**f**4) Central part of intervertebral disc shows the nucleus pulposus with a few persisting notochordal cells and endochondral bone from adjacent vertebrae above and below. The growth cartilage of the surface end plates stains *dark purple* and the fibrocartilage of the disc stains lighter *purple*. Note the absence of the bone within the physeal cartilage from the nucleus pulposus to the metaphyseal bone of vertebrae. [*Pink* staining acellular material between the nucleus pulposus and disc cartilage is a fibrous remnant, not bone.] Marker length = 50 μm. (**f**5) Periphery of the intervertebral disc (annulus fibrosus) is seen at the right and disc itself at the left. Adjacent vertebral bone is seen above and below. Note the well-organized laminar arrangement of the fibroelastic tissue from vertebra to vertebra. Marker length = 100 μm

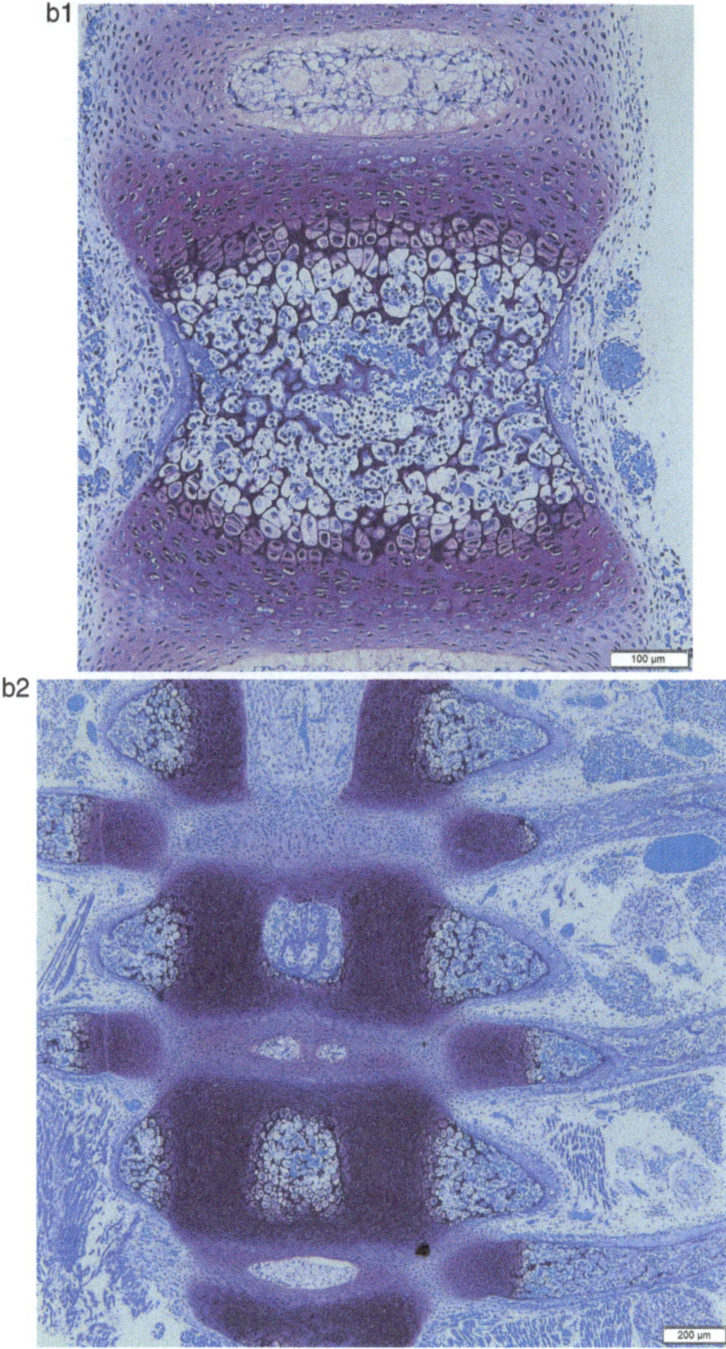

Fig. 3.7 (continued)

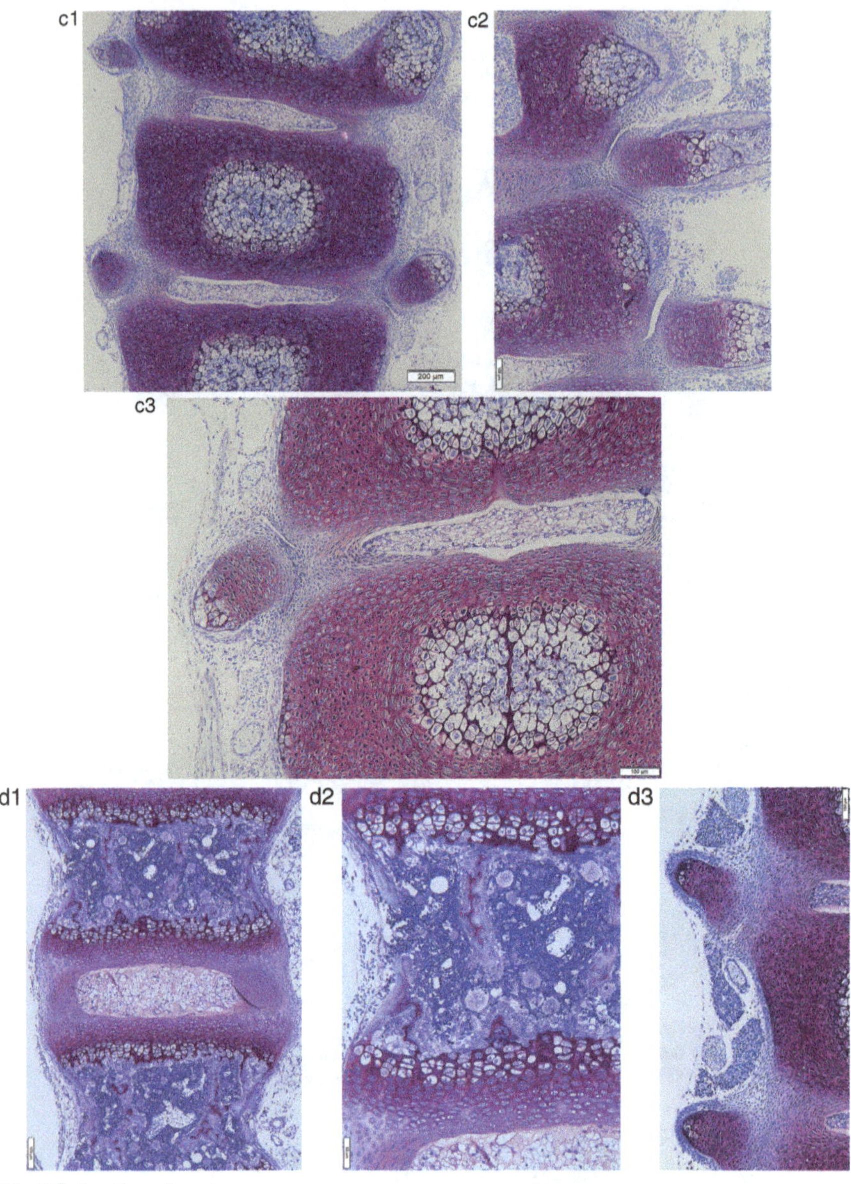

Fig. 3.7 (continued)

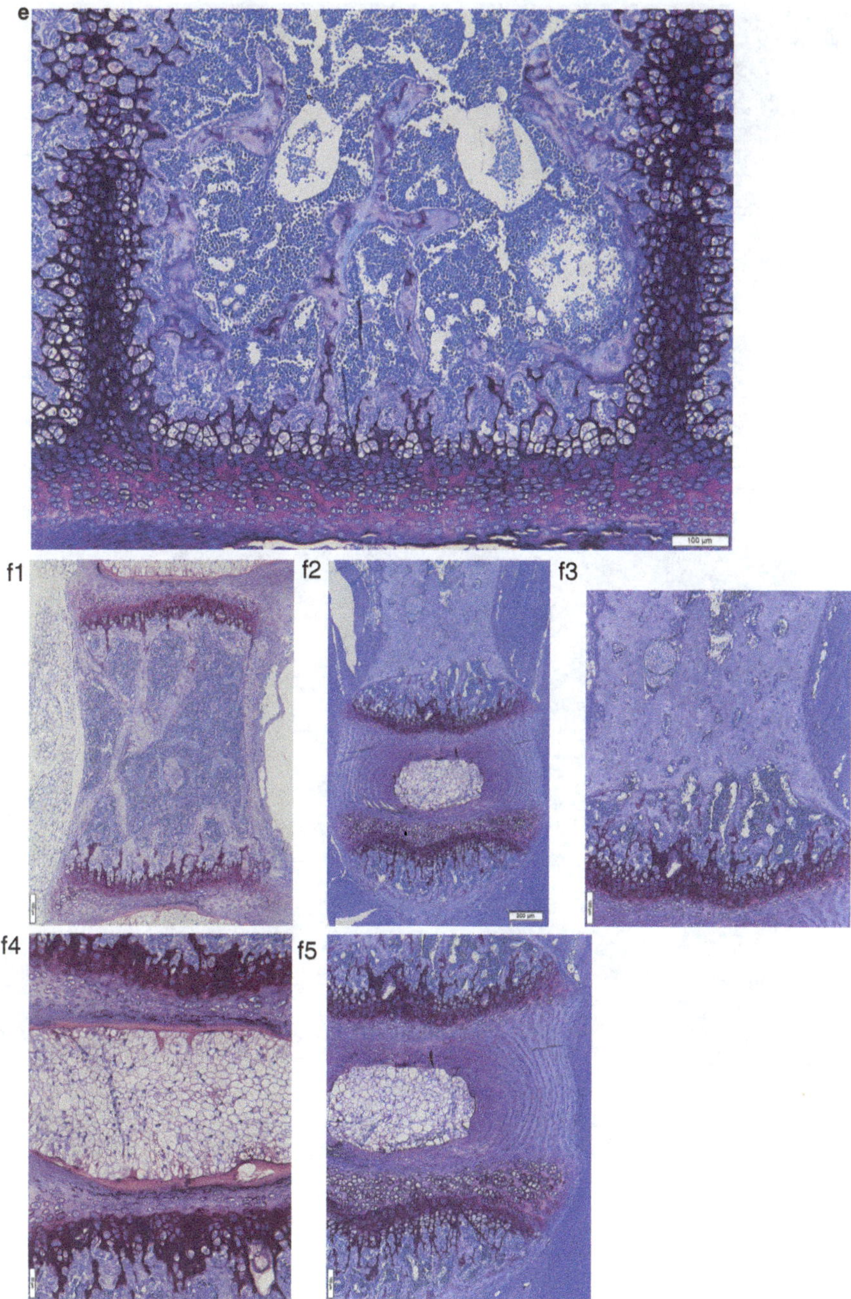

Fig. 3.7 (continued)

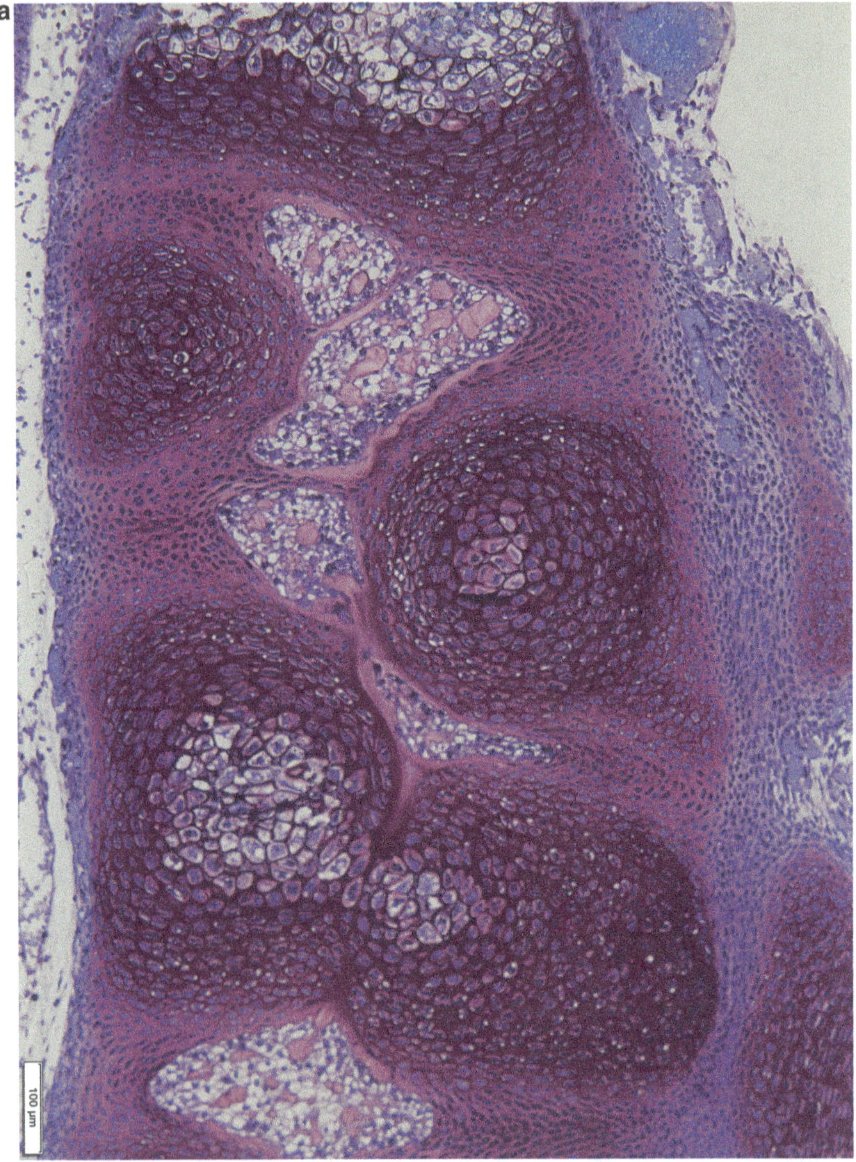

Fig. 3.8 Histologic sections of pudgy (pu/pu) mouse vertebrae demonstrate the markedly abnormal structural development. All sections were embedded in JB4 plastic and stained with 1 % toluidine blue. (**a**). Embryo (#58). Thoracic vertebrae show marked disorganization of vertebral body and intervertebral disc formation. The intervertebral discs (nucleus pulposus) are extremely irregular in shape, size, and position with orientation in transverse, vertical, and oblique planes. There are two paravertebral masses of the cartilage, the right side, separated from the vertebral bodies by a cellular (abnormal interzone) region. Ribs (fused and asymmetric) will be originating from this mass from planes not seen at this level of sectioning. Marker length = 100 μm. (**b**1). Embryo (#58). Additional developmental irregularity is seen. A paravertebral cartilage mass

3.4.2 Serial Section Assessments

Serial sections were done in 11 pudgy mice: pu/pu # 61 (embryo); #s 24, 47, and 56 (newborn); # 50 (3 days); # 15 (9 days); #s 27 and 29 (3 weeks); # 13 (6 weeks); and #s 35 and 45 (4 weeks). Serial sections were also done in two non-affected mice: pu/+ # 63 (embryo) and # 25 (newborn). Assessment of serial sections clearly defines the sequence of structural changes across the entire extent of the vertebral column and allows for better interpretation of the changes, especially in malformed and irregular structures. While not as numerous, many sections per segment were also made from other specimens.

Fig. 3.8 (continued) extending the length of two vertebral bodies and an interposed cellular region is seen. Marker length = 200 μm. (**b**2). Embryo. A higher magnification view from the top half of (**b**1) highlights the paravertebral cartilage regions from which abnormal ribs originate. These regions are separated from the adjacent vertebral bodies by a cellular tissue consistent with an abnormal joint interzone indicating a probable pseudo-synovial cavity formation. Marker length = 100 μm. (**b**3) A higher magnification view from the lower half of (**b**1) shows relatively normal rib development even though the adjacent vertebrae and discs are markedly abnormal. Marker length = 100 μm. (**c**) This photomicrograph is from embryo (#59). The lumbar region demonstrates bifid vertebra formation with involuting notochord intervening between the two segments. The intervertebral discs are irregular in shape, size, and position and are often incomplete across the width of the spinal segment. Marker length = 100 μm. (**d**) Photomicrographs from another pudgy embryo (#61) are shown. (**d**1) Midsagittal section shows developing vertebral bodies but markedly abnormal nucleus pulposus which are almost continuous at multiple levels. Marker length = 100 μm. (**d**2) Section from the same block of the tissue shows irregular development but markedly different patterns of the nucleus pulposus and vertebral body shape and position. The spinal cord is at the left and large vessels at the right. Marker length = 100 μm. (**d**3) Series of six black and white photomicrographs cut in the parasagittal plane spans the same longitudinal region of the developing vertebral axis but includes sections passing from one side to the other. These sections were chosen from a serial section series of several dozen. This shows the variable changes at the same level from side to side. (**e**) Newborn pudgy mouse (#24) photomicrographs are shown. (**e**1) Markedly abnormal intervertebral disc and vertebral body formation in the lumbar region is demonstrated in a newborn pudgy mouse. Note the relationships of irregularly shaped vertebral bodies to each segment of the malpositioned nucleus pulposus (notochordal origin). The cranial end of the vertebral column is at the top and the caudal end at the bottom. Marker length = 100 μm. (**e**2) A montage merges three consecutive longitudinal images to highlight the extensiveness and histologic variability of abnormalities throughout the vertebral column. Each of the three merged images has a marker length = 200 μm. (**e**3–**e**5) Higher power photomicrographs demonstrate each section of the montage from (**e**2) in greater detail from top (cranial) to bottom regions of thoracolumbar vertebrae. Marker length for each image = 100 μm. (**f**1–**f**5) Higher magnification photomicrographs show an even higher level of cell patterning detail in sections from the montage progressively from top to bottom. Marker length = 100 μm in each image. (**g**1–**g**3) Three higher magnification photomicrographs from a different newborn pudgy mouse (#47) further highlight the abnormalities. Marker length = 100 μm in each image. (**h**1, **h**2) Adjacent photomicrographs from a 3-week-old pudgy mouse show how bone formation defines the final abnormal shapes. Marker length = 200 μm. (**i**) Photomicrograph demonstrates a section of thoracolumbar spine from a 4-week-old pudgy mouse. Marker length = 200 μm

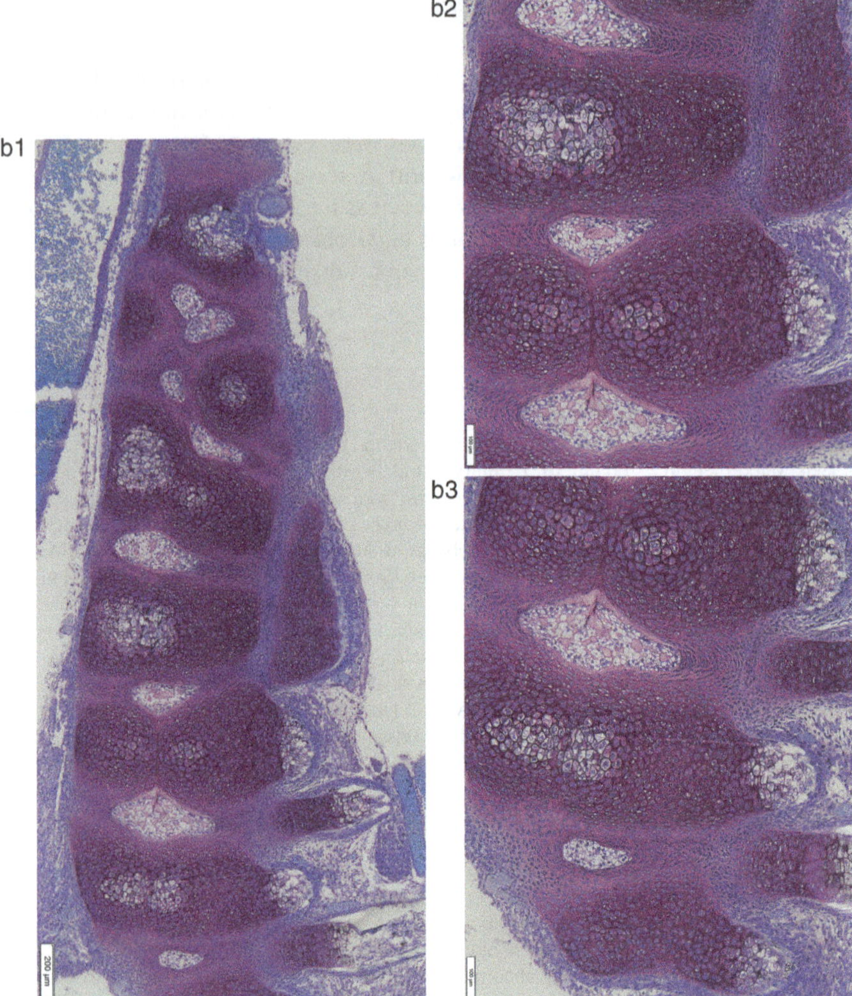

Fig. 3.8 (continued)

3.4.2.1 Development of Vertebrae

Normal Vertebral Body and Intervertebral Disc Development

Pu/+ 63, Embryo

Normal vertebral column segmentation is seen with alternating vertebral bodies and intervertebral discs. The cartilage bodies show large central collections of hypertrophic chondrocytes as the endochondral sequence begins to form the vertebral body bony centra (Fig. 3.7a).

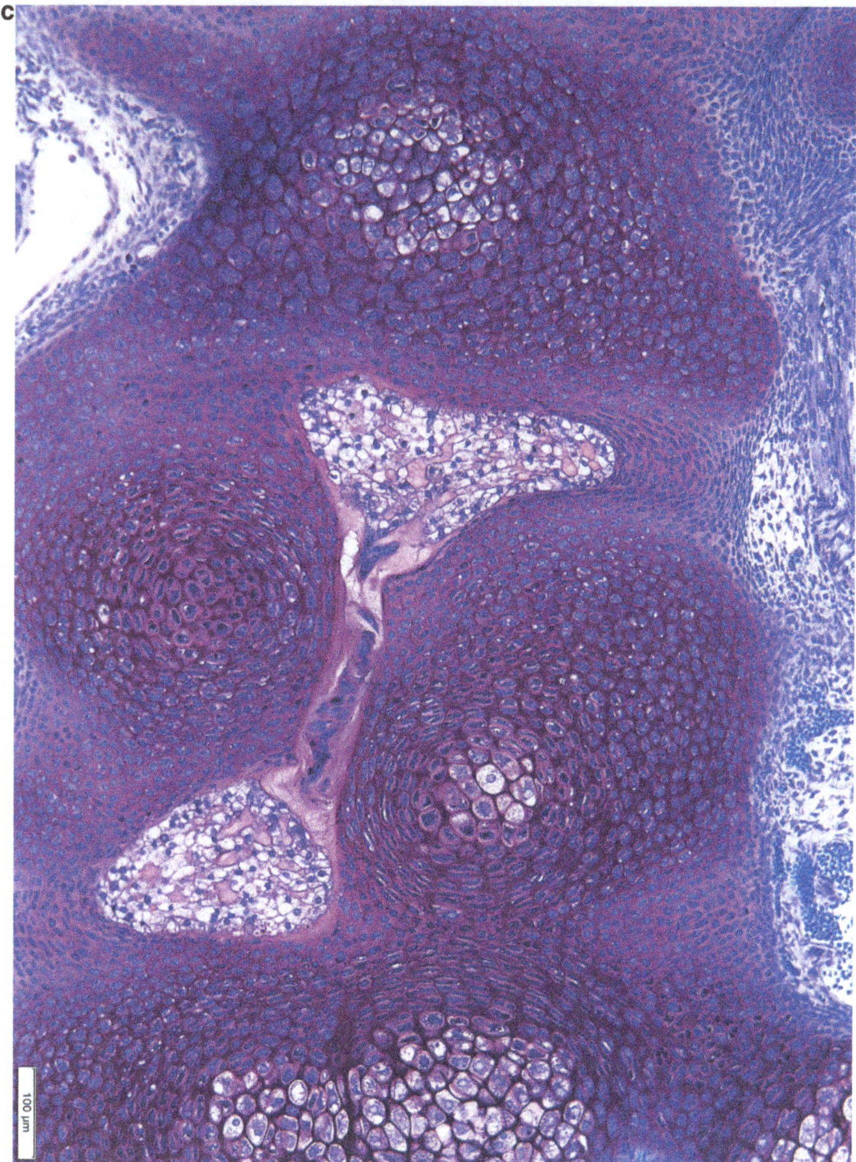

Fig. 3.8 (continued)

Pu/+ 25, Newborn

Serial sections of the vertebral bodies and discs were made from ventral to dorsal (171 sections). Coronal cuts through the vertebral bodies show bone tissue to be forming by the endochondral mechanism. Once the vertebral body reaches a certain size, the walls undergo bone formation by the intramembranous mechanism. A

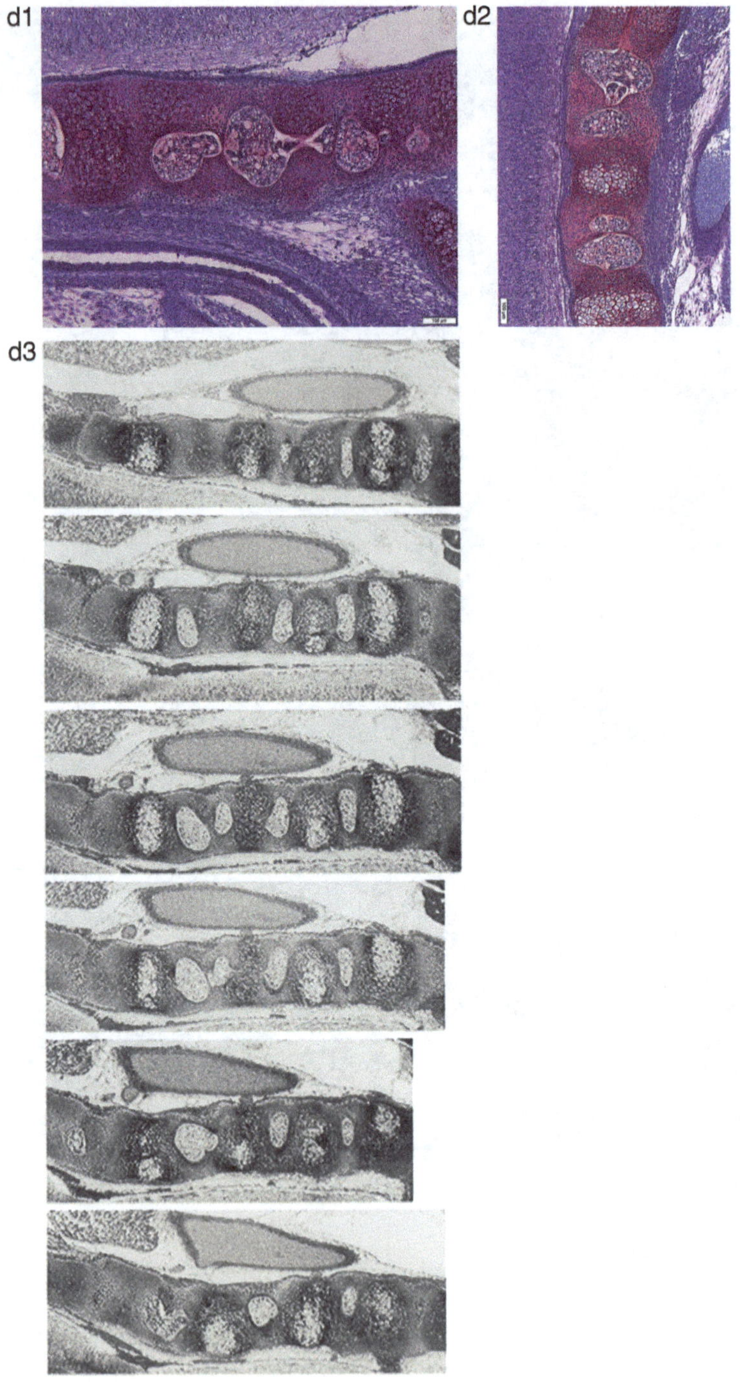

Fig. 3.8 (continued)

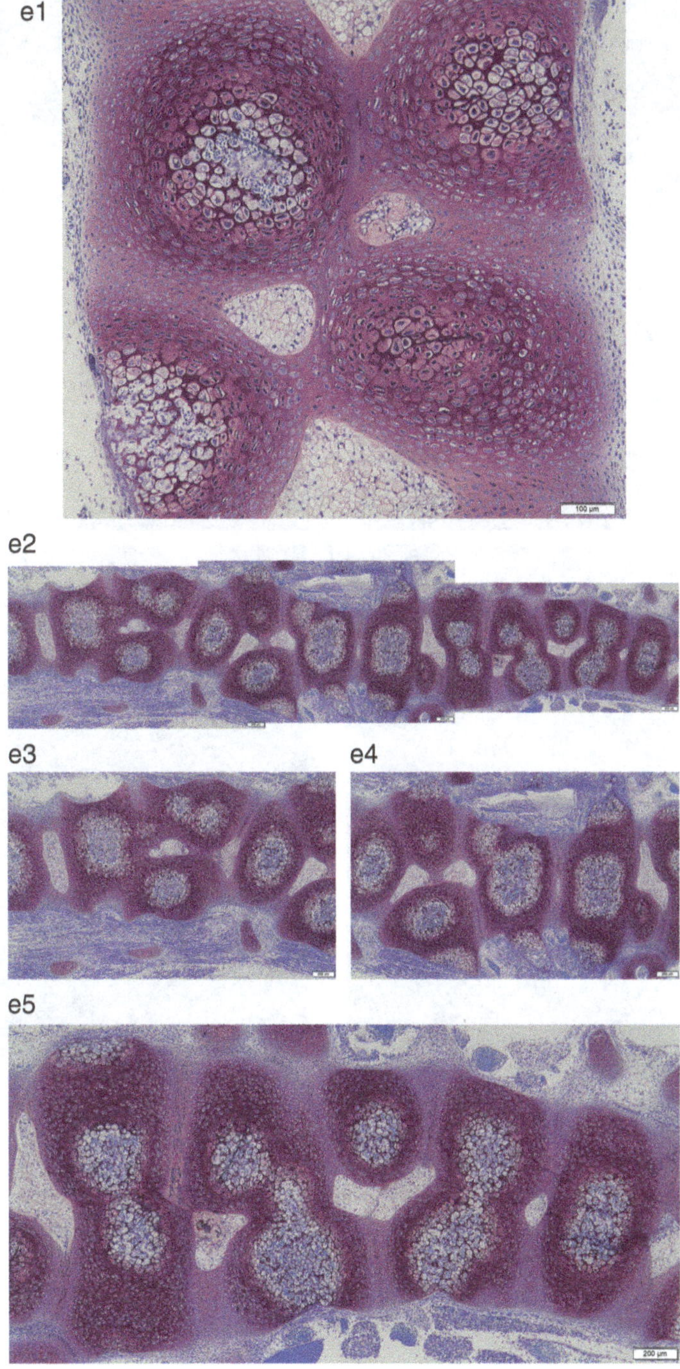

Fig. 3.8 (continued)

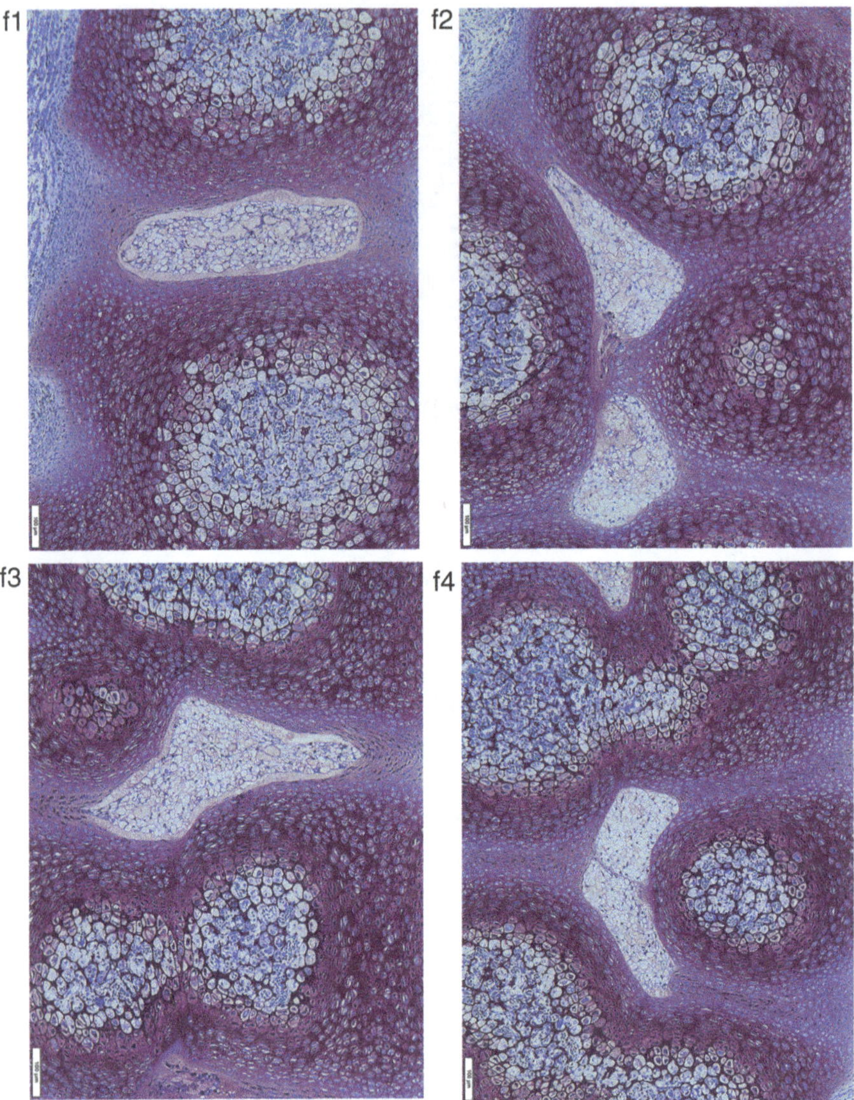

Fig. 3.8 (continued)

growth plate has been established at the cranial and caudal ends of each developing vertebral body. There is marrow cavitation with extensive osteoclastic resorption of bone and cartilage cores within the vertebral body. The disc regions have formed well but still have a significant cartilage component with the fibrous component (annulus fibrosus) at the periphery. The notochordal cell remnants are seen in the nucleus pulposus (Fig. 3.7b1, b2).

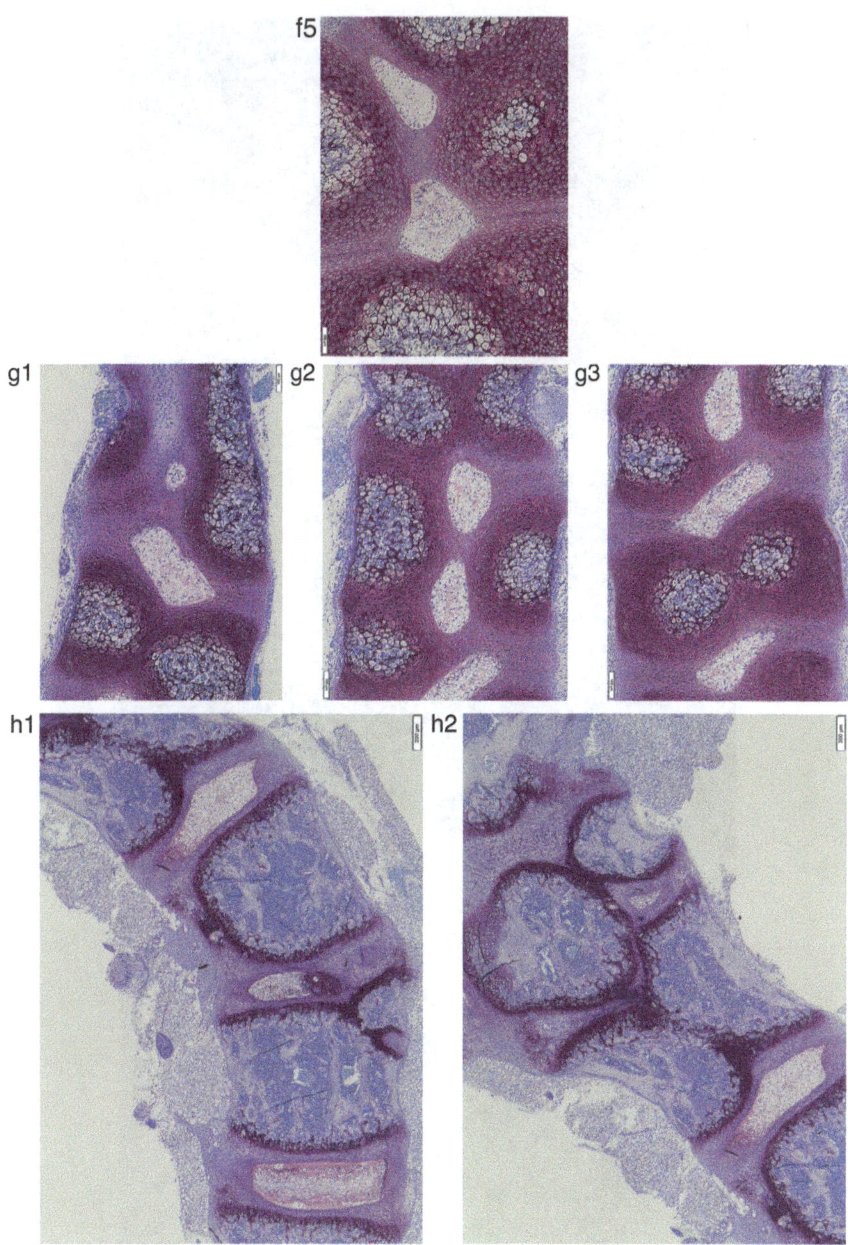

Fig. 3.8 (continued)

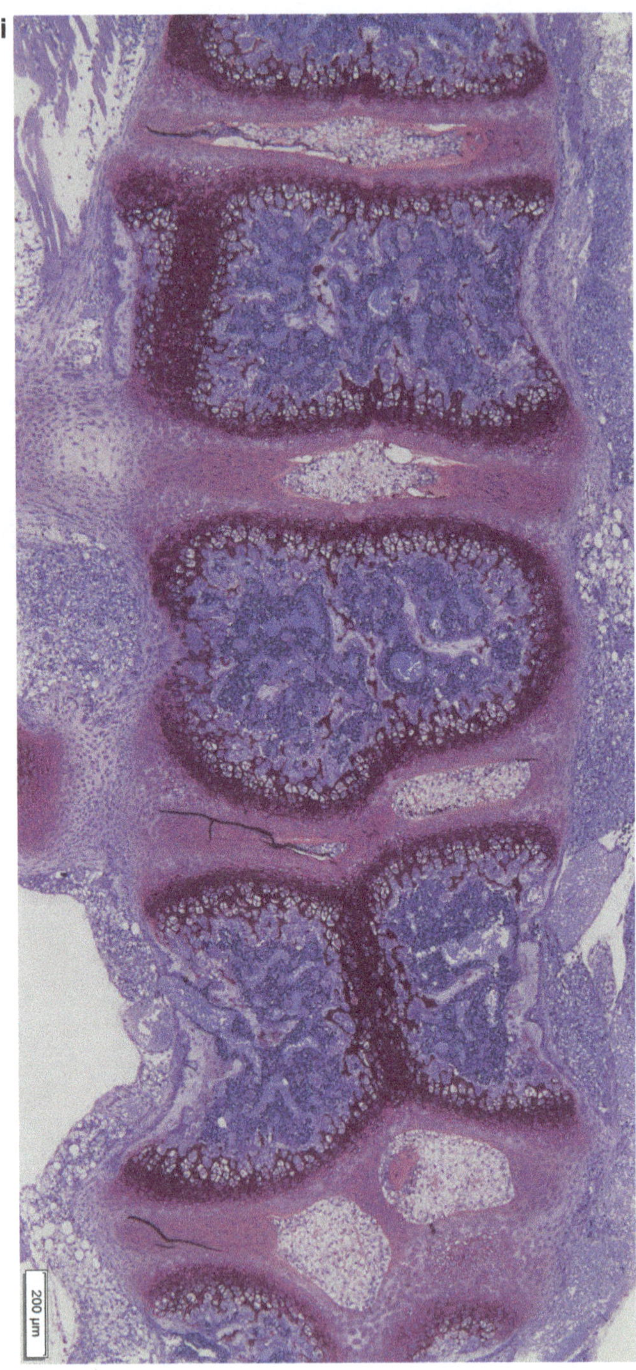

Fig. 3.8 (continued)

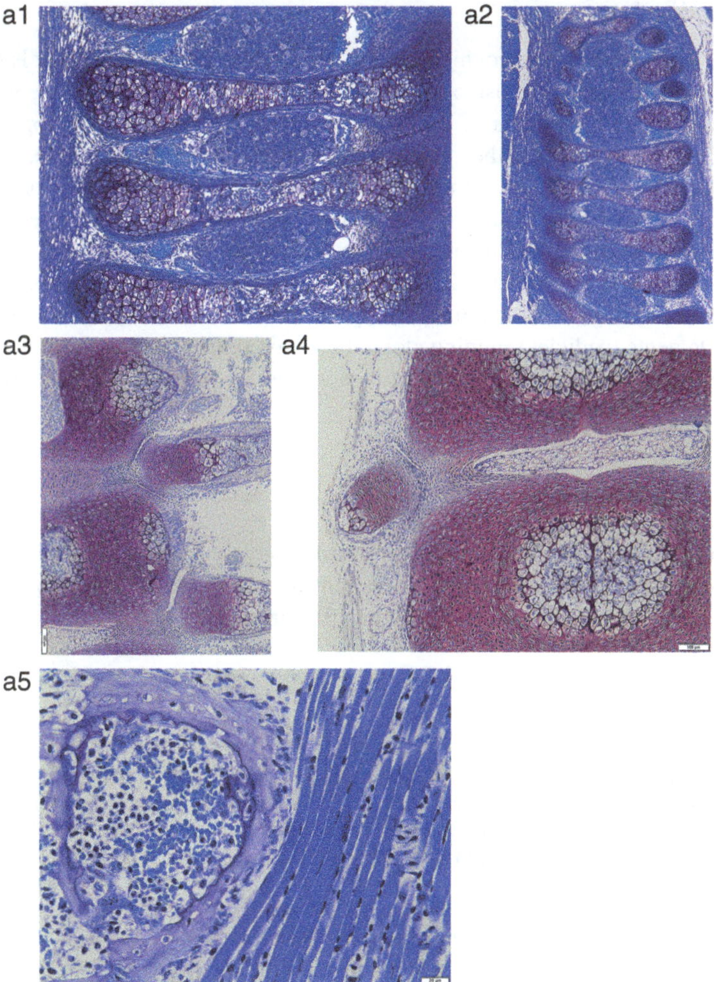

Fig. 3.9 Normal non-affected (pu/+) mouse and abnormal pudgy (pu/pu) mouse rib formation is illustrated. All sections in this figure were embedded in JB4 plastic and stained with 1 % toluidine blue. (**a**1–**a**5) Normal rib formation. [Normal rib formation is also seen adjacent to the vertebrae and intervertebral discs in Fig. 3.7.] (**a**1) Normal rib formation in the embryo (#63) is illustrated. Multiple ribs are shown sectioned along their long axes posteriorly adjacent to the vertebrae. Cartilage development at either end is by the endochondral mechanism but the cortices of the shafts are formed, as in long bones, by the periosteal intramembranous bone. Typical ganglionic tissue is seen between four ribs. (**a**2) In this photomicrograph there is a continuous rib at the top and five at the bottom but the plane of sectioning has moved posterior to the rib shafts in between. The ganglionic tissue above is seen but at this stage and site appears continuous, although fine fibrous septae are present between each ganglion. Below the ganglia appear segmented lying between the ribs. (**a**3) Newborn rib formation is demonstrated with beginning synovial joint cavitation seen. Marker length = 100 μm. (**a**4) Another costo-vertebral site in newborn mouse is seen. Marker length = 100 μm. (**a**5) Newborn normal mouse shows rib shaft in cross section undergoing periosteal intramembranous bone formation. Internally, marrow cavitation has occurred and last remnants of the calcified cartilage against the inner cortex are being resorbed. The adjacent muscle is seen in longitudinal section. Marker length = 20 μm. (**b**1–**b**13) demonstrate

Pu/+ 18, Newborn

In some sections the hypertrophic zone of the vertebral body bone center is present in a 360° arc. Considerable osteoclast resorption of the cartilage within the vertebral body is seen. Cell remnants of the notochord persist within the developing disc. Ribs are seen adjacent to the cranial border of the thoracic vertebrae. In some instances a joint cavity is seen but more often the differentiating rib/vertebral mesenchymal cell collections are in continuity. In sections that are more dorsal in the body (closer to the spinal canal), the ossification center, signified by hypertrophic chondrocytes, of the lateral elements (transverse processes) is seen. Separate areas of the cartilage (neurocentral cartilage) persist between the vertebral body and lateral elements (pedicles) running along the longitudinal axis perpendicular to and continuous with the growth plate cartilage at both surfaces of the vertebral body (Fig. 3.7c1–c3).

Pu/+ 21, One Day

In some sections, ganglion tissues are noted to be continuous over two or three segments, although high-power observation always shows a continuous fibrous septum between adjacent ganglia. The ganglia are regularly spaced.

Fig. 3.9 (continued) abnormal pudgy rib formation. (**b**1). Embryo (#59). A cross section of ribs adjacent to the spinal cord shows markedly irregular shapes due to fusions and origins from the paravertebral cartilage accumulations. Irregular shaped and continuous collections of ganglionic cells are seen at the top. Marker length = 100 μm. (**b**2). Embryo (#61). A large array of asymmetric ribs (in cross and oblique sections) is seen; irregularly shaped, continuous appearing ganglionic collections are interspersed although these are separated by thin fibrous septae. Marker length = 100 μm. (**b**3–**b**9). Embryo (#62). Each of these sections shows the abnormal, asymmetric ribs and interspersed abnormal (continuous, misshapen) ganglionic tissue from pudgy embryo sections. Fused and branching ribs are seen as are the abnormal paravertebral cartilage tissue accumulations from which these fused ribs originate. There always appear to be cellular collections between the cartilage growth regions of the abnormal fused/asymmetric rubs that we interpret as attempts at joint formation such that there is no direct cartilage continuity between vertebral body and abnormal rib. Marker lengths for (**b**4) = 200 μm and (**b**5–**b**9) = 100 μm. (**b**10). Newborn. Several vertebral levels are seen with markedly abnormal but differing rib formation. At the upper right, a normal rib is forming. At the mid-right there is a continuous loop joining rib origin sites in a paravertebral cartilage tissue mass from which fused ribs will originate. At the upper left, the forming rib from the paravertebral cartilage accumulation is adjacent to two vertebrae, the upper site normal but the lower also relating to the sidewall of the vertebra via an interposed joint interzone type of cellularity. At the lower left, there is a broad rib origin with the paravertebral mass encompassing three adjacent vertebral bodies. Marker length = 100 μm. (**b**11) Higher-power view of rib origin from the right side, middle region. Marker length = 100 μm. (**b**12, **b**13) Bone structure of branching/fused ribs from 3-day-old. Although the patterning is very abnormal, the fused/branching ribs develop from the paravertebral cartilage using the endochondral growth mechanism, and the shafts form using the normal intramembranous bone mechanism. Marker lengths = 100 μm

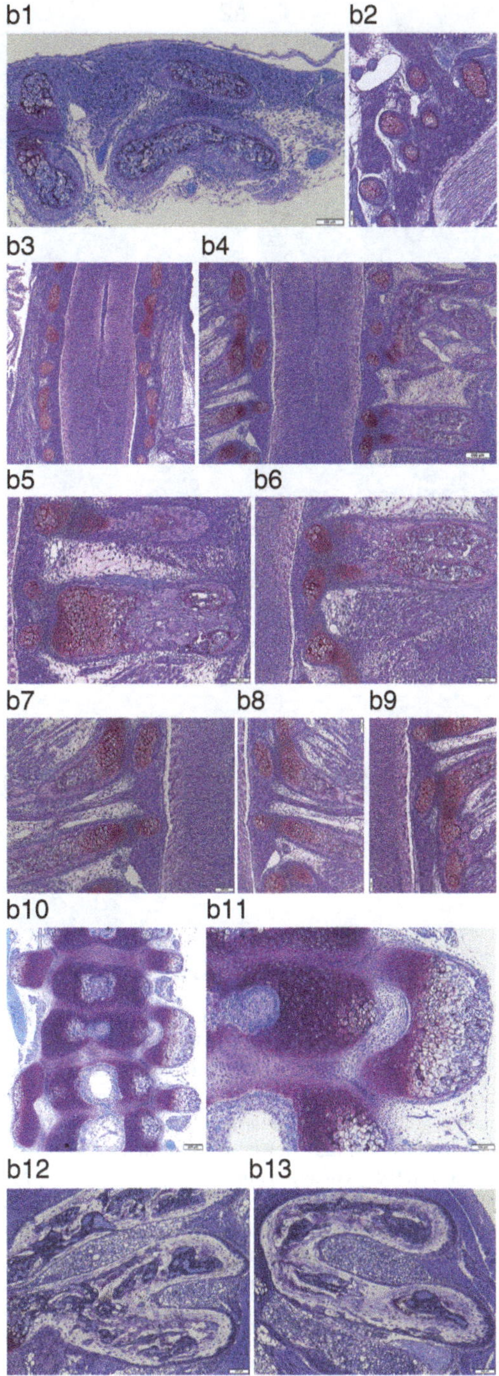

Fig. 3.9 (continued)

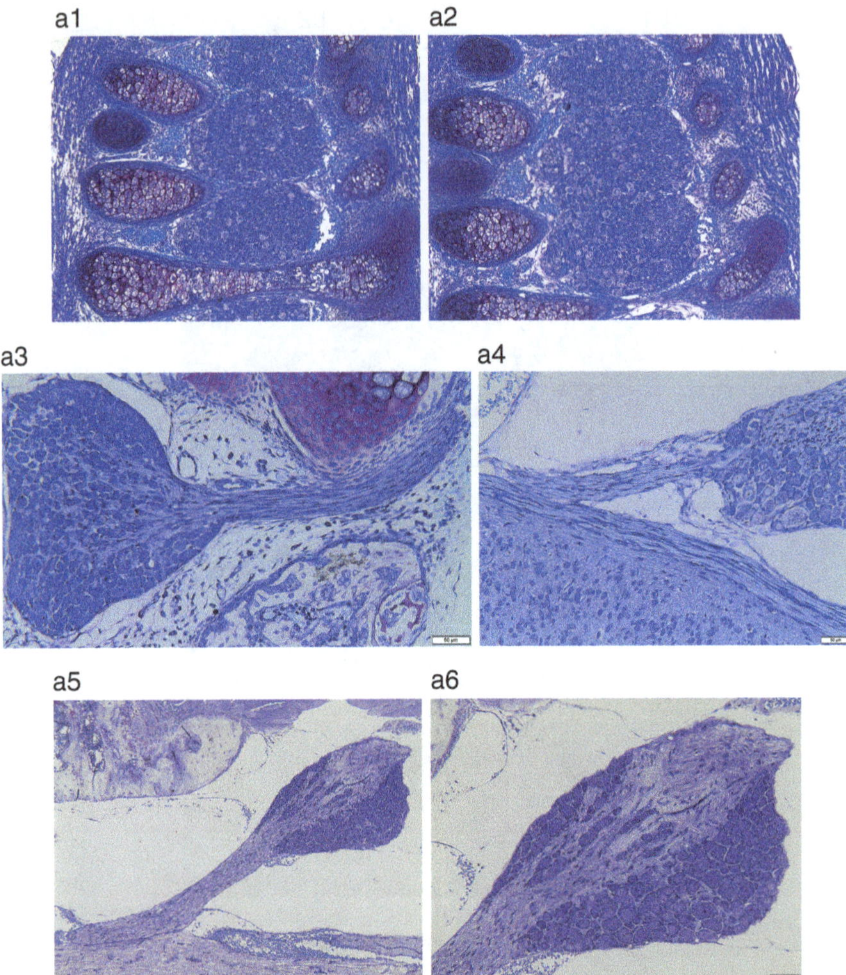

Fig. 3.10 Normal non-affected (pu/+) mouse and pudgy (pu/pu) mouse ganglia formation is shown. All sections in this figure were embedded in JB4 plastic and stained with 1 % toluidine blue. (**a**1–**a**2) Normal ganglia formation [additional examples of normal ganglia formation are also seen in (**a**1, **a**2)]. Embryo. As part of the same sequence of sections shown for rib development in (**a**1, **a**2), there is continuous ganglionic normal tissue present although thin fibrous septae separate the immediately adjacent collections. (**a**3) Newborn. A normal ganglion adjacent to ribs is seen. Marker length = 50 μm. (**a**4) At 9 days of age, a normal ganglion is shown. Marker length = 50 μm. (**a**5, **a**6). Normal ganglia are seen from a mouse at 6 weeks of age. Marker lengths = 100 μm (**a**5) and 50 μm (**a**6). (**b**1–**b**3) Abnormal pudgy mouse ganglion formation [abnormal ganglia are also seen in relation to abnormal pudgy rib formation in Fig. 3.9b]. Abnormal-appearing ganglia are seen in relation to abnormal costo-vertebral formation in each histologic section from different embryonic mice. The ganglia are misshapen and appear continuous although thin septae are seen between adjacent collections. At least some of the misshapen structure appears due to relating to abnormally shaped ribs. Marker lengths = 100 μm

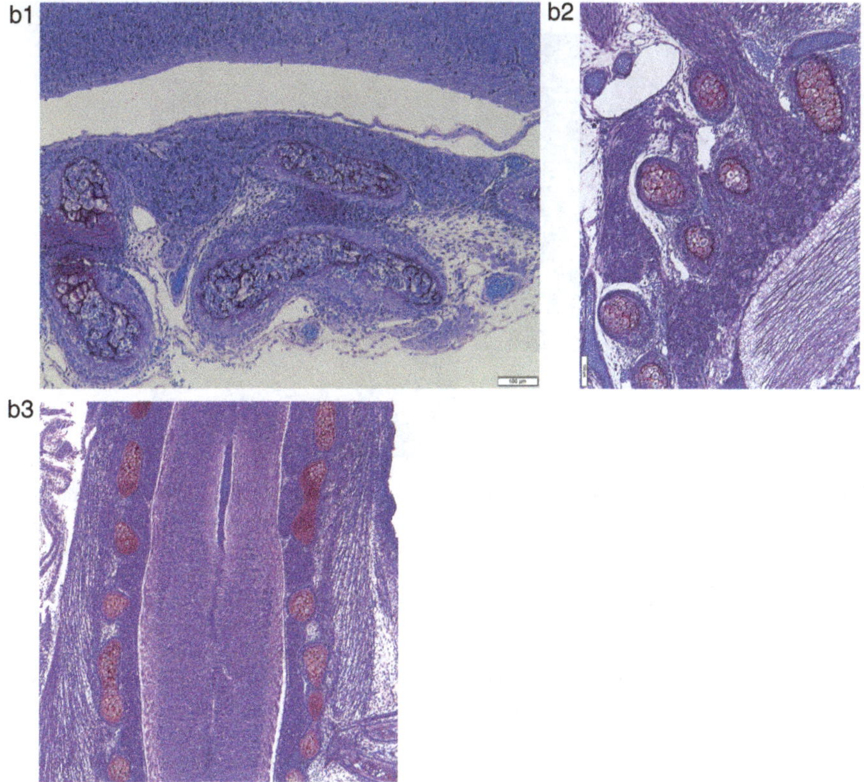

Fig. 3.10 (continued)

Pu/+ 49, Three Days

The bipolar neurocentral growth plate between the vertebral body and neural arch bone centers (pedicle) has been well established.

Pu/+ 44, Six Days

The vertebral bodies have reached a state of development showing superior and inferior surface growth plates, endochondral growth with marrow cavitation, side walls of vertebral body formed by intramembranous ossification, and well-formed intervertebral discs with the central nucleus pulposus and peripheral annulus fibrosus (Fig. 3.7d1, d2).

Pu/+ 14, Nine Days

The growth plates at either end of the vertebral bodies are now well developed. The neurocentral cartilage physes are well formed. The sidewalls of the vertebral bodies are composed of an intramembranous bone cortex. There is extensive marrow cavitation internally (Fig. 3.7e).

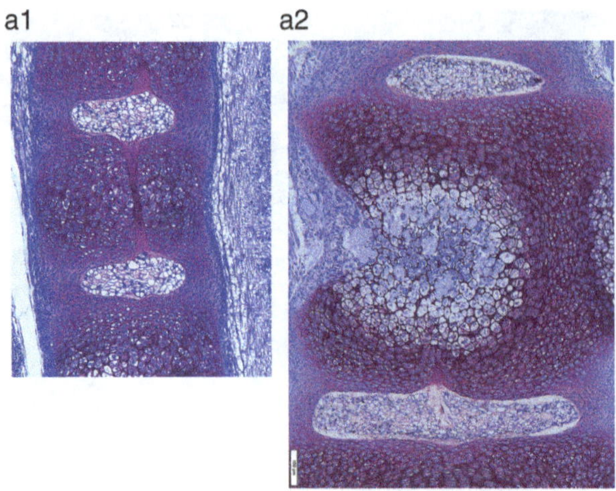

Fig. 3.11 Normal non-affected (pu/+) mouse and pudgy (pu/pu) mouse intervertebral disc formation. All sections were embedded in JB4 plastic and stained with 1 % toluidine blue. (**a**1) Normal intervertebral disc formation. [Normal intervertebral disc formation is also illustrated throughout Fig. 3.7.] Embryo. The nucleus pulposus is highly cellular and still shows fibrous septae passing from vertebral bodies above and below along the line of its absorption from its earlier continuous notochord stage. The annulus fibrosus region is highly cellular with the cartilage tissue the main component of the fibrocartilaginous composition. The cells peripherally however are beginning to align in the curvilinear lamellar fashion from the vertebral body above to the vertebral body below. (**a**2) In a non-affected mouse at 4 h of age, the nucleus pulposus is still cellular but there is beginning fibrous tissue conversion of the annulus. Marker length = 100 μm. (**a**3) At 3 weeks there is a marked decrease in cellularity of the nucleus pulposus. The annulus fibrosus is increasingly fibrous peripherally and the characteristic lamellar orientation of the collagenous tissue is evident. Marker length = 100 μm. (**a**4) At 3 weeks, the tissue is shown from the center of the intervertebral disc to the periphery. The nucleus pulposus shows diminished cellularity from the embryonic period. The annulus fibrosus also shows diminished cellularity and more structured orientation of the fibrous lamellae. Marker length = 100 μm. (**a**5) At the periphery of the intervertebral disc, the nucleus pulposus is no longer seen and the annulus fibrosus predominates across the entire disc. Note the regular orientation of the laminae. Marker length = 100 μm. (**a**6) At the outermost periphery of the annulus fibrosus, the plane of sectioning shows the characteristic well-organized cross-hatched orientation of the fibrous lamellae. Note the viable appearance of the fibroblasts that are still an integral component of the tissue. Marker length = 100 μm. (**b**1, **b**2) Pudgy (pu/pu) mouse intervertebral disc formation. [Abnormal pudgy intervertebral disc formation is also seen in all parts of Fig. 3.8.]. (**b**1) [Section oriented with cranial direction at the left and caudal at the right.] At 9 days, there are many intervertebral disc irregularities. At the right, there is no disc tissue between two adjacent misshapen vertebral bodies. Wherever this situation occurs, there are always flattened chondrocytes between the two adjacent models of the cartilaginous vertebral bodies. Other regions show markedly misshapen discs that can lie in oblique or vertical planes, be round to bulbous in shape, either thinner or wider, and incomplete across the width of the column. Marker length = 100 μm. (**b**2) Similar disc abnormalities in a 3-day-old mouse are seen. Marker length = 100 μm. (**b**3) The annulus fibrosus at the outermost periphery of a misshapen disc is also misshapen with the cross-hatched intersecting lamellae very irregular in orientation. Marker length = 100 μm

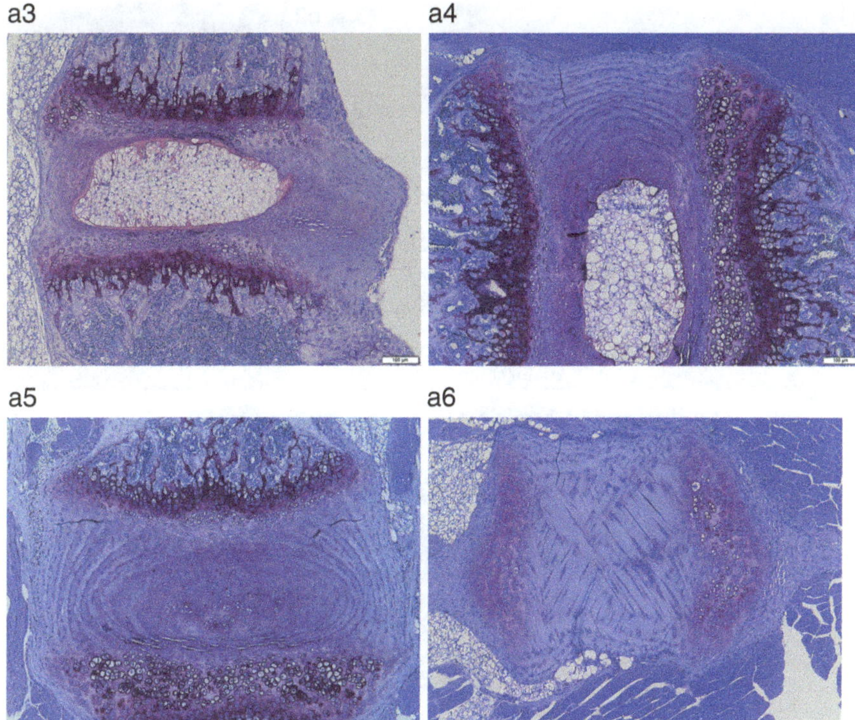

Fig. 3.11 (continued)

Pu/+ 11, 26, 28, Three Weeks

Each of the various elements maintains its relative positioning with increase in size and tissue differentiation occurring. Excellent maintenance of the developmental sequence is seen. Growth plates at upper and lower ends of the vertebral bodies are active in forming the endochondral bone. The lateral walls of the vertebral bodies are composed of the cortical bone synthesized by the intramembranous mechanism (Fig. 3.7f1–f5).

Pu/+ 38, Four Weeks

The normal developmental growth structure persists.

Pu/+ 12, Six Weeks

The growth plates at upper and lower margins of the vertebral body are thinner and less active with the minimal cartilage persisting in the "metaphyseal" regions toward each vertebral bone body.

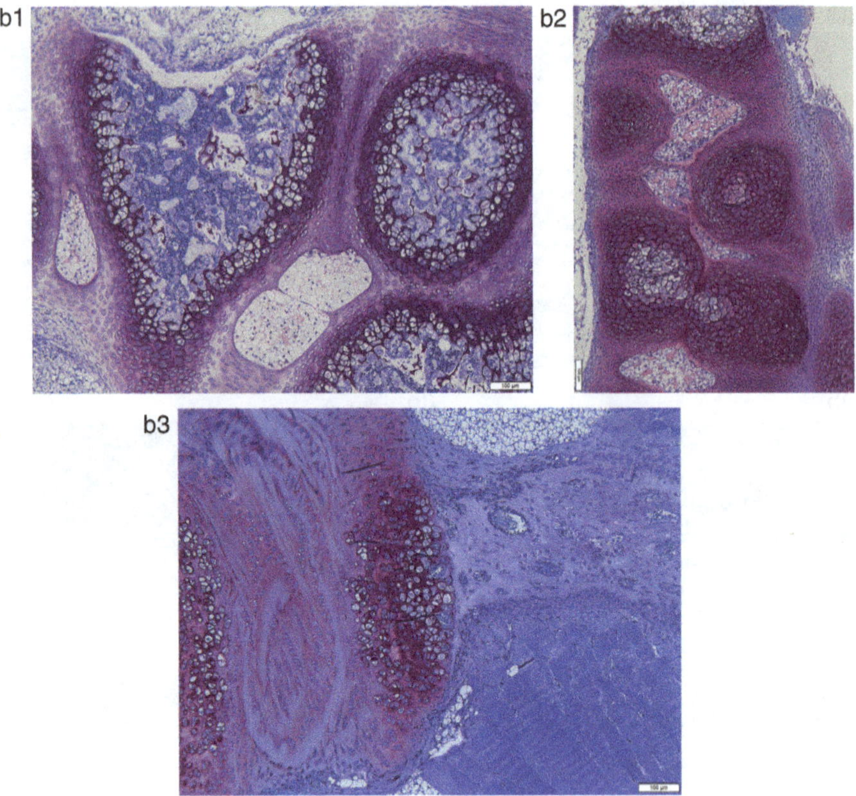

Fig. 3.11 (continued)

It is evident in all histologic sections, from six days of age onward, showing the superior and inferior vertebral body cartilage growth regions immediately adjacent to the intervertebral discs that no secondary ossification center bone forms in the growth cartilage between the disc and the growth plate for vertebral body longitudinal growth. In other words, there are no linear collections of bone tissue to correspond with the linear radiodensities of these regions seen on specimen radiographs.

Pudgy Mouse Vertebral and Intervertebral Disc Development

Overview

The structural abnormalities of the individual vertebrae and ribs and of the intervertebral discs are due to abnormal pattern formation. Bone tissue of the vertebrae and ribs develops normally by endochondral and intramembranous mechanisms but in inappropriate shapes and positions because of the abnormal cartilage models. The major abnormalities of the ribs are at their proximal ends adjacent to the vertebrae. In certain regions, the cartilage from two adjacent vertebrae is contiguous without interposition of the nucleus pulposus or annulus fibrosus. At these sites we never observe complete cartilage model continuity from misshapen vertebra to misshapen vertebra but rather an interface of flattened chondrocytes from each model against one another. The intervertebral discs are invariably abnormal adjacent to abnormally shaped or positioned vertebral bodies showing (i) thicker or thinner accumulations; (ii) vertical or oblique positioning; (iii) abnormal shaping into round, hexagonal, curvilinear, or other patterns; (iv) discontinuity to varying extents across the area separating two adjacent vertebral bodies; and/or (v) complete absence between adjacent (and malformed) vertebral bodies. It is the histopathology that gives the best understanding of the irregularities associated with the disorder.

The range of abnormalities is illustrated in great detail in Fig. 3.8a–i and the accompanying figure legends. Three affected pudgy mice from the same litter (sibling group # 8) assessed at the late embryo stage show the marked variability from mouse to mouse. These are siblings #58, seen in Fig. 3.8a, b1–b3; # 59 in Fig. 3.8c; and # 61 in Fig. 3.8d1–d3. In Fig. 3.8a, a histologic section of several malformed vertebrae and intervertebral discs from the embryo #58 provides an overview of the markedly deranged developmental process. Further findings in embryos are shown in Fig. 3.8b–d. Figure 3.8b1 shows a different segment of the spine than Fig. 3.8a, b2, b3 are the upper and lower sections of Fig. 3.8b1, magnified to highlight cell detail. Figure 3.8c from embryo #59 shows further marked abnormality. Figure 3.8d1, d2 from embryo #61 highlights the marked intervertebral disc abnormalities, while Fig. 3.8d3 shows the variability in structure across the same width of the tissue. Figure 3.8e, f is from newborn pudgy mouse #24 with multiple views highlighting the complexity of the irregular process. Figure 3.8e1 shows the lumbar irregularities. Figure 3.8e2 is a montage of three adjacent sections to highlight the multilevel changes, while Fig. 3.8e3, e4, e5 magnifies for detail the findings from the top, middle, and bottom segments of the montage in Fig. 3.8e2. Further findings in newborn mouse #24 are shown in Fig. 3.8f1–f5. They support one of the main conclusions of this study: *the more irregular the shape and position of the intervertebral discs, the more irregular are the shape and position of the adjacent vertebral bodies.* Figure 3.8g1–g3 shows abnormalities in newborn mouse #47; Fig. 3.8h1, h2 in 3-week-old mouse # 29; and Fig. 3.8i in 4-week-old mouse # 45.

3.4.2.2 Development of Ribs

Normal Rib Development (pu/+)

The normal rib develops by endochondral and intramembranous ossification mechanisms. The segment of the rib immediately adjacent to the vertebrae (referred to as the dorsal/vertebral/proximal segment) develops as a cartilage model forming a characteristic costo-vertebral synovial joint. The rib articulates with the two adjacent vertebral bodies at two facets of the costo-vertebral joint. In the embryo this joint forms, as in the larger appendicular extremity joints, from a continuous mass of initially undifferentiated mesenchymal cells. As the articular surfaces start to be shaped, there is a gradual resorption of the interzone cells as the synovial cavity forms. In some newborn, and at one week in all, the costo-vertebral synovial joints are seen. The rib then continues growth from a single growth plate relating to the shaft of a single rib. The cortex of the rib, as in the long bone or the eventual walls of the vertebral body, forms by the periosteal intramembranous bone with a marrow cavity an integral part of its structure. The characteristic oblique intercostal skeletal muscle layers link adjacent ribs. At their distal ends, seven ribs merge with the sternal cartilage and six terminate as free-floating unattached structures. Normal rib development is illustrated from embryo (# 63) in Fig. 3.9a1, a2, from newborn (# 18) in Fig. 3.9a3, a4, and from newborn (# 25) in Fig. 3.9a5. It can also be seen in Fig. 3.7 adjacent to normal vertebral and intervertebral disc development.

Pudgy Rib Development (pu/pu)

The abnormal rib development in the pudgy mouse originates, or is centered, in an abnormal paravertebral cartilage accumulation or mass. Instead of a single symmetric rib origin, the pudgy mouse rib cage is often characterized by having two or more (and as many as five or six) ribs originating from the same cartilage mass. There are also regions where no cartilage mass forms, leading to an absent or deleted rib, and there is often asymmetric positioning of the paravertebral cartilage mass, both in relation to right and left sides and cranially/caudally leading to increased or decreased spacing between either individual or fused ribs on one side. Some ribs develop in the normal position in relation to two adjacent vertebral bodies. A synovial costo-vertebral joint usually forms, although it may be misshapen and in abnormal position. Even when the ribs just lateral to the costo-vertebral articulations are fused, the costo-vertebral regions tend to form although the individual joints may be slightly widened. There are generally a decreased number of individual ribs; increased numbers are not seen. In these abnormal situations, individual widened physeal cartilage directing rib formation may form just lateral to the paravertebral mass and then bifurcating or even trifurcating into separate shafts or fused ribs may continue from that mass for a variable distance. This paravertebral mass is sometimes separate from the lateral vertebral bodies (in the form of a false fibrous joint or pseudarthrosis) but sometimes shows full bone continuity from the vertebral body to paravertebral longitudinal mass to ribs. Histology shows normal endochondral or intramembranous bone formation at the cellular level, even though the position and patterning is abnormal. Abnormal rib development is highlighted in Fig. 3.9b1–b13 with studies illustrated from embryos (#s 59, 61, 62 [sibling group #8]), newborn (# 24), and 3 days of age (# 50) pudgy mice.

3.4.2.3 Development of Ganglia

Normal Ganglia Formation (pu/+)

A large mass of segmentally contiguous highly cellular tissue along the longitudinal axis slightly displaced from the vertebrae characterizes normal dorsal root ganglia development. While the characteristic appearance of the well-developed ganglion is of a circular to oval separate cellular mass from which fibrils pass to link with the spinal cord, initial development is from a much larger cellular mass as characterized below. When histologic sectioning is made in the coronal plane through the more anterior parts of the vertebral column, the ganglia appear as separate accumulations, lying between the ribs as they develop. With serial sectioning, however, and moving in a more dorsal direction at initial observation, the ganglia appear almost as a continuous cellular mass. The close contiguity of each is readily apparent in the newborn. Observation at high power, however, shows that each ganglion is separated from the next by a thin fibrous septum. With time, the individualized separate nature of each ganglion and its connecting fibrils to the spinal cord is evident. By 4–6 weeks of age, the ganglion is about half cellular with the rest fibrillar and continuous with the root that approaches the spinal cord. Normal ganglia formation is illustrated in Fig. 3.10a1–a6 from embryo (# 63), newborn (# 53), 9-day-old (# 14), and 6-week-old (# 12) pu/+ mice.

Pudgy Mouse Ganglia Formation (pu/pu)

Many sections, from newborn animals, show extensive seemingly continuous collections of ganglia almost as prominent as the vertebral bodies and adjacent ribs. These connect to the spinal cord via the normal-appearing roots but have irregular shapes. When examined at high power however, they do appear separate, although often only by very thin fibrous septa. The abnormal shape could be considered as at least partially secondary to the abnormal immediately adjacent vertebral body and rib shapes. Examples of these histologic findings are shown in the photomicrographs in Fig. 3.10b1–b4 from embryos (#s 59, 61, and 62) and adjacent to abnormal ribs in Fig. 3.9b.

3.4.2.4 Development of Intervertebral Discs

A general overview of intervertebral disc development was provided in section **a)** above but additional information is presented here.

Normal Intervertebral Disc Formation (pu/+)

The intervertebral discs are originally derived from the sclerotomes. The central nucleus pulposus contains remnants of the notochord with some surviving cells. At the late embryonic time periods, remnants of the involuting notochord can be seen still within the midline of the vertebral bodies as thin fibrous bands. At later time periods, the notochordal tissue is limited to the annulus fibrosus of the intervertebral disc. Peripherally the annulus fibrosus is composed of layers of fibrous tissue arrayed in well-organized cross-hatched fashion for structural support. The fibrous tissue of the intervertebral disc adjacent to the cranial and caudal surfaces of the vertebral bodies merges with the physeal cartilage of those surfaces of the body. There is no bone formation between the physis and the purely fibrous tissue of the intervertebral disc. Linear radiodensities at this level seen on radiographs are indicative of a thin layer of calcification, but histological examination of numerous sections at varying ages shows no bone formation. Besides the images of normal formation in Figs. 3.7 and 3.9a, additional illustrations are presented in Fig. 3.11a1–a6 from embryo (# 63), 4 h (#8), and 3 weeks (#s 26 and 28).

Pudgy Intervertebral Disc Formation (pu/pu)

The intervertebral discs are markedly abnormal in the pudgy mouse. Their structure approaches the normal only when positioned between two vertebral bodies that are only minimally misshapen. At all other levels of the pudgy spine, the discs are abnormal. These abnormalities have been described in the sections on radiology and histology above. The outer layers of the annulus fibrosus demonstrate markedly irregular cross-hatching compared to the well-patterned cross-hatching alignment in the normal. Abnormal intervertebral disc development is highlighted in Figs. 3.8, 3.9b, and 3.11b1–b3.

3.5 Computerized Three-Dimensional Reconstructions of Histologic Sections

Studies were done in two affected pudgy mice (pu/pu) [one embryo (#61) and one newborn (#47)] and one normal mouse (pu/+) [embryo (#63)]. Color-coded tissue components show the cartilage, nucleus pulposus, annulus fibrosus, and bone shaping abnormalities both alone and in relation to each other in a way not appreciated by whole mount preparations, thin histologic sections, or radiographs. These reconstructions can be assessed at positions throughout a 360° rotational axis. Examples of tissue pattern assessed by the three-dimensional reconstructions of serial histologic sections are illustrated in Fig. 3.12a–e.

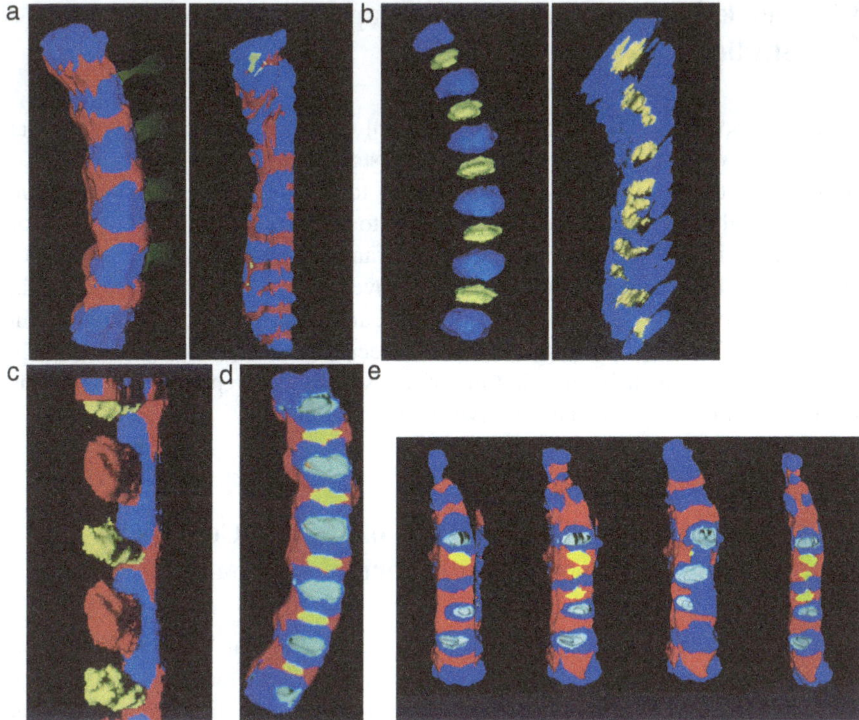

Fig. 3.12 Three-dimensional computerized reconstructions of serial histology sections of vertebral columns in embryonic normal non-affected (pu/+) mice and pudgy (pu/pu) mice are shown. The color codings are slightly different in some studies but are explained for each. (**a**) Normal segmentation is seen on the left with *blue* representing the cartilaginous vertebral body and *red* the annulus fibrosus of the disc. Linear *green* color outlines the spinous processes. The study in an age-matched pudgy mouse is shown at the right. Note the markedly abnormal segmentation in the pudgy specimen with *blue* representing the cartilaginous vertebral bodies and *red* the annulus fibrosus. Some of the vertebral bodies are combined as single masses, the annulus fibrosus is often incomplete at several levels across the width of the column, and at the *top yellow* represents peripherally placed nucleus pulposus of the discs where the annulus is lacking. (**b**) Normal segmentation is shown in the midsagittal plane at the left as alternating *blue* and *yellow* regions. The *blue* in this image outlines the bone of the developing vertebral body and *yellow* the nucleus pulposus of the disc. A pudgy mouse vertebral column is shown at the right tracing the same parameters. In this *blue* and *yellow* image, the lack of normal periodicity is seen clearly. *Blue* represents the bony centers of the pudgy column, while *yellow* shows the irregularly shaped and positioned intervertebral discs. (**c**) Sections of four contiguous vertebrae are seen. *Blue* represents the cartilage model of the vertebral body and, at the right, *red* is the annulus fibrosus. The three vertebrae to the left represent a three-fourth cutaway that removed most of the cartilage of the vertebral bodies and the annulus fibrosus. The *blue* of the remaining cartilage vertebral body is at the bottom, the vertebral body bone (*circular* to *oval*) is *red*, the nucleus pulposus of the intervertebral disc is *yellow*, and the persisting part of the annulus fibrosus at the bottom is also *red*. (**d**) Another reconstruction of a normal embryo spinal segment shows the annulus fibrosus in *red*, cartilage model of the vertebral body *dark blue*, ossification in the center of the vertebral body *light blue*, and intervertebral disc *yellow*. (**e**) Four abnormal pudgy spines from embryonic and newborn pudgy mice, in varying rotations, are projected. *Color code*: annulus fibrosus, *red*; cartilage model of vertebral body, *dark blue*; vertebral body bone, *light blue*; and intervertebral disc, *yellow*. Multiple abnormal variations are seen

3.6 Chick Embryo Vertebral Development (Previous Studies)

We have reported our investigations on normal chick vertebral development that involved (1) whole mount preparations, (2) histologic and three-dimensional histochemical studies of embryos from three to nineteen days of incubation, (3) sequential ^{3}H-thymidine autoradiography to demonstrate shifting patterns of DNA synthesis, and (4) endochondral and intramembranous mechanisms characteristic of vertebral bone formation [13]. An overview of findings is presented in Fig. 3.13, whole mount images in Fig. 3.13a–c, and histology photomicrographs in Fig. 3.13d–k. These are presented briefly here because of the similarity in very early structural development before the time of assessment in the pudgy mouse in this report (which begins in the late embryo stage).

3.7 Radiology and Histopathology of Human Congenital Scoliosis of the Spine (With Normal Comparisons)

A male child with congenital scoliosis with many vertebral body, intervertebral disc, and rib irregularities, born at 38 weeks to a diabetic mother, died at nine days of age of cardiorespiratory failure. This material has been published previously but the spinal specimen radiograph and histopathology are presented here to compare with the remarkably similar pudgy mouse findings [7]. The dorsal bony elements were present but irregular on radiologic examination from the ninth thoracic to the third lumbar vertebrae. Shortened, fused, and absent ribs (nine ribs on the right, ten ribs on the left), Sprengel's deformity (elevated scapula), unilateral clubfoot, and hypoplastic ilia were present. The clinical neuromuscular examination was normal. At autopsy, there was no diastematomyelia, spina bifida, or meningomyelocele, and neuropathology studies revealed no gross or histologic abnormalities of the spinal cord or peripheral nerves. Associated abnormalities involved the lungs (hypoplastic), genitourinary system (bilateral renal hypoplasia, hydronephrosis, hydroureter, double right ureter, hypoplastic prostate), and heart (patent ductus arteriosus and double aortic arch).

3.7.1 Normal Fetal and Neonatal Human Vertebral Bodies

In specimens from an approximately 18-week-old fetus and a 14-day-old neonate who died of encephalitis, the cartilage growth plates participating in endochondral ossification are present at the cranial and caudal regions of each vertebral body (Fig. 3.14a–c). These are continuous with the intervertebral disc tissue and are not structured as epiphyses with secondary ossification centers. In the fetal spine (cut in

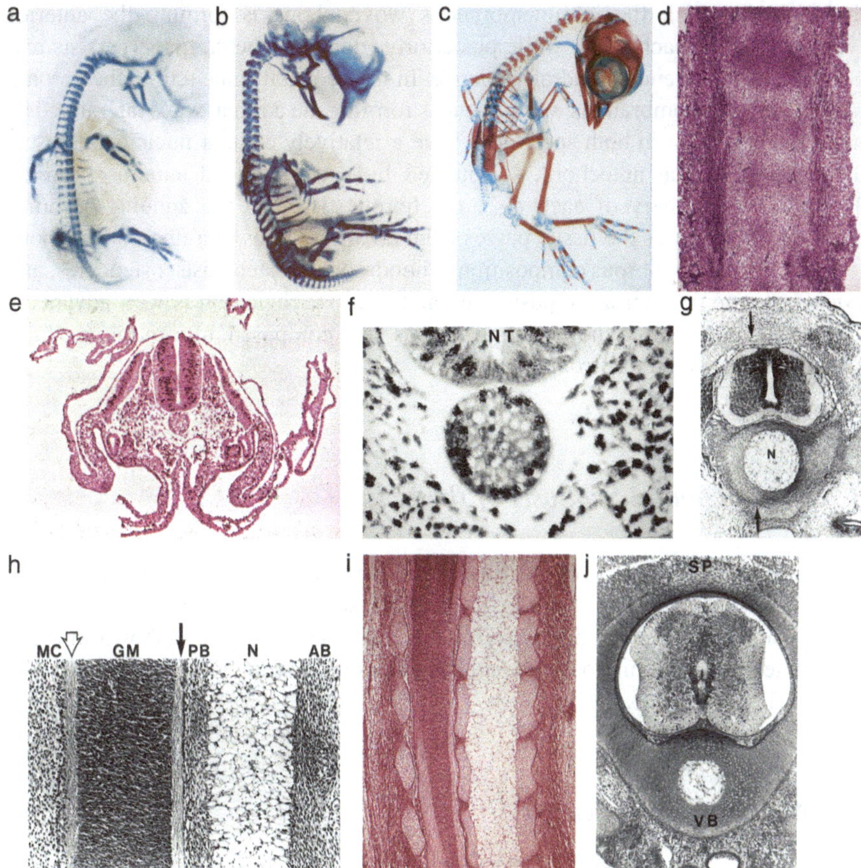

Fig. 3.13 Images from chick embryo vertebral development are shown. (**a**, **b**, **c**) Whole mount sections show progressive changes with the development from all blue staining when the skeleton is preformed in the cartilage; to some red staining as initial bone formation occurs to almost entirely red staining as bone formation predominates with blue staining persisting at the ends of the bones only due to articular cartilage and growth plates. Beginning at the left, chick embryos are (**a**) 6½ days (stage 30), (**b**) 7½ days (stage 32), and (**c**), 14 days (stage 40). [There are no secondary ossification centers in chick long bones (with rare exceptions) unlike in the mouse and human.] (**d–J**) Histologic sections show varying tissue development in early axial formation at 3 days and at 12 days of development at the cartilage model stage of vertebral body formation. Cross-sectional and longitudinal views are shown. All sections were embedded in paraffin and stained with hematoxylin and eosin. (**d**) Alternating light and dark segmentation regarding cell density is shown. At the right and left sides, the linear vertical collections are the myotomes. (**e**, **f**) Autoradiographs of transverse section of chick embryo at 3 days (stage 22) show active cellular uptake (*black dots*) following injection of egg with 3H-thymidine. *NT* neural tube, *N* notochord. Photomicrographs at 5 days (stage 27). Cells of the neural tube, notochord, and adjacent sclerotomes are active. (**g**) Transverse section shows the spinal cord and notochord surrounded by differentiating cartilage destined to form the vertebral body. *Arrows* indicate the parasagittal plane of sectioning just off the midline axis for the photomicrograph in (**h**). *N* notochord. (**h**) Midline axial structures along the axis outlined by *arrows* in (**g**) are seen from anterior to posterior as follows: *AB* anterior (ventral) vertebral body, *N* notochord, *PB* posterior (dorsal) vertebral body,

the sagittal plane), the intramembranous woven bone is forming the anterior (ventral) wall of each body, while posteriorly (dorsally) the cartilage persists and participates in the endochondral sequence. In the neonatal spine (cut in the coronal plane), the intramembranous woven bone is forming the cortical walls laterally. The intervertebral discs in both specimens have a relatively cellular nucleus pulposus, the remnant of the notochord, surrounded by a well-defined annulus fibrosus. Toward the periphery of each disc, the fibrous tissue in the annulus becomes more predominant as the tissue passes from a cartilaginous to a fibrocartilaginous to an almost totally fibrous composition. Chondrocytes, osteoblasts, osteocytes, and osteoclasts are appropriately positioned and marrow cavitation is well advanced. The trabeculae within the vertebral bodies are composed of central acellular cartilage cores and the surrounding bone.

3.7.2 Congenital Scoliosis of the Spine

Assessments of the same segment using different techniques included a detailed drawing of the congenital scoliosis of the spine, specimen radiograph, light microscopic paraffin-embedded hematoxylin and eosin-stained section (shown in black and white), and light microscopic paraffin-embedded safranin O-fast green-stained section (Fig. 3.15).

3.7.2.1 Radiologic Appearance

Anteroposterior radiographs of the spine demonstrated the nature and extent of the congenital scoliosis deformity including the abnormal ribs. The specimen radiograph, shown in Fig. 3.15, focuses on the vertebral column.

Fig. 3.13 (continued) dark arrow ventral white matter of the spinal cord, *GM* gray matter (neuroepithelium) of the spinal cord, *open arrow* dorsal white matter of the spinal cord, *MC* mesenchymal cells in the region which will form the dorsal neural arch and spinous process. (**i**) Hematoxylin and eosin-stained section along the same plane as outlined in (**g**), at a slightly later stage showing vertebral body cartilage differentiation. (**j**) Transverse section is seen at 9 days (stage 35) showing the cartilage surrounding both the notochord and spinal cord. *VB* vertebral body, *SP* dorsal spinous process. There is diminished cellularity and decrease in size of the notochord, relative to the developing vertebral body. (**k**) Chart outlines the developmental sequence, in Hamilton-Hamburger (HH) stages, for the chick embryo from day 3 to day 19 for whole mount, histology, and autoradiography [(**a–c**) and (**e–k**) reprinted with permission from: Shapiro, F. Vertebral development of the chick embryo from 3-19 of incubation. J Morphol 1992; 213: 317-333. Wiley-Liss]

k

Temporal outline of chick vertebral development[*]

Age (days)	H–H stage	Whole-mount embryos	Histological appearance	Autoradiography Dt	Mt	St	SC	Nt	VB	NA	SP
3	22	Somites clearly seen	Neural tube (spinal cord) and notochord prominent; dermatome, myotome, sclerotome well defined	++	++	++	++	++	—	—	—
5	27	Alternating areas of light and dark stain but no true vertebral definition	Mesenchymal cells surround spinal cord laterally and notochord	———	———	———	++	++	++	—	—
6	29	Individual blue staining vertebrae[1]	Cartilage well differentiated in vertebral body						↓		
6½	30	Vertebrae more deeply stained; dorsal spinous processes defined							↓		
7	31	Increasing size of vertebrae; all blue staining persists	Vertebral bodies, neural arches, and spinous processes now all formed in cartilage				++	+	++	++	++
8	34	↓					++	0	++	++	++
9	35	↓	Cartilage cell hypertrophy in vertebral body				+	0	++	++	
10	36	↓	Considerable decrease in size of notochord				+	0	+[3]	+	+
11	37	↓	Cartilage cell hypertrophy in vertebral body increases				+	0	+[3]	+	+
12	38	↓									
13	39	Vertebral body bone formation seen at ventral and dorsal surfaces as indicated by red stain[2]	Periosteum synthesizing intramembranous bone at ventral surface of vertebral body; vascular invasion of vertebral body hypertrophic chondrocytes; separate centers of hypertrophic chondrocytes in neural arches				+	0	+[3]	+[3]	+
13½	40	Bone formation increases throughout vertebral bodies into neural arch									
15	41	↓	Increased bone formation throughout; in some transverse sections of body, notochord no longer present				+	0	+[3]	+[3]	+[3]
16	42	Vertebrae red staining throughout indicative of bone formation, except at intervertebral regions; proximodistal gradient seen; vertebrae red in cervical, thoracic, and upper lumbar areas but lower lumbar, sacral vertebrae still stain primarily blue	↓						↓		
17	43	Increasing bone differentiation caudally	Bone formation continues; remnants of notochord in vertebral bodies only; growth in vertebral bodies at cranial and caudal surfaces						↓		
19	45	↓									

[*]H–H, Hamburger-Hamilton stage; Dt, dermatome; Mt, myotome; St, sclerotome; SC, spinal cord; Nt, notochord; VB, vertebral body; NA, neural arch; SP, spinous process; ++, heavily labeled; +, labeled; 0, no labeling; —, structure has not yet appeared.
[1]Blue staining = uptake of alcian blue by cartilage tissue; [2]red staining = uptake of alizarin red by bone tissue; [3]no^{3}H-thymidine labeling in hypertrophic chondrocytes.

Fig. 3.13 (continued)

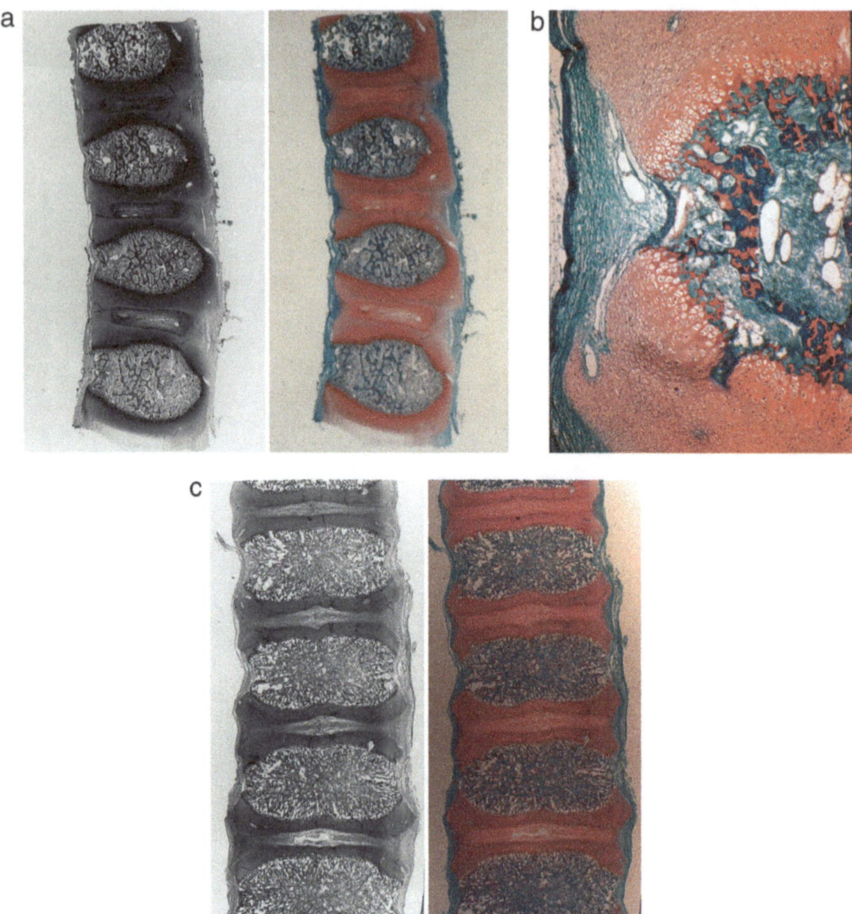

Fig. 3.14 Histologic sections of normal human lumbar vertebrae from embryonic (18 week) and postnatal (14 days) time periods are shown. The tissue sections were embedded in paraffin. (**a**). In the embryonic period, vertebral body and intervertebral disc formations are shown. A *black* and *white* photomicrograph (hematoxylin and eosin stained) is at the left and a safranin O-fast green stain at the right. The tissue has been sectioned in the sagittal plane with the anterior (ventral) surfaces at the left. (**b**) A higher magnification of the anterior vertebral body is shown from the embryonic period (sagittal plane, anterior (ventral) surface at the left, safranin O-fast green stain). (**c**) In the postnatal period, the vertebral body bone has replaced most of the cartilage model. Growth plates at the upper and lower body surfaces are seen. A black and white photomicrograph (H and E preparation) is at the left and safranin O-fast green preparation at the right. The tissue was sectioned in the coronal plane

3.7.2.2 Histopathology

Figures 3.15 and 3.16a–c illustrate the morphologic findings. The thoracic 7 (T7) and T8 vertebrae are normal. Endochondral ossification at the site of continuing longitudinal growth occurs at the cranial and caudal regions of the

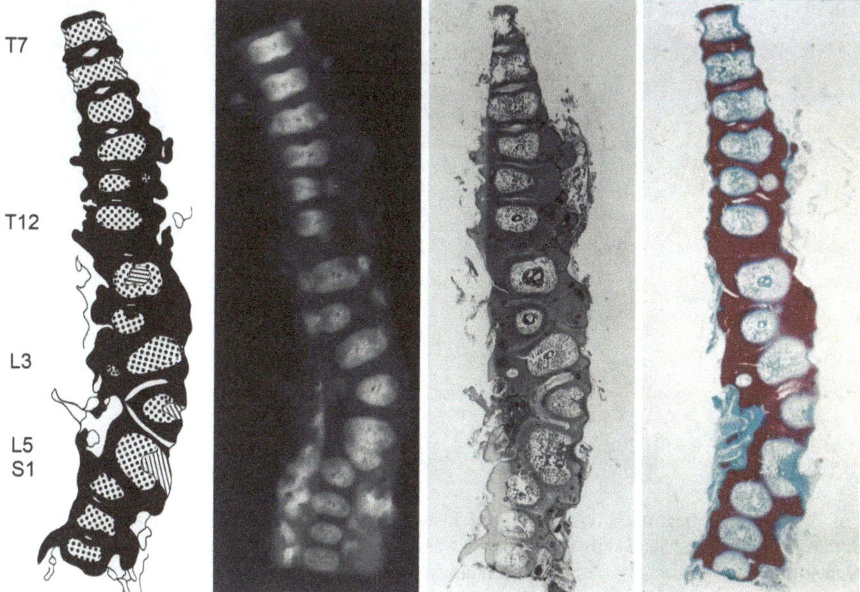

Fig. 3.15 Human congenital scoliosis. A large segment of the spine from the seventh thoracic vertebra to the third sacral vertebra (T7-S3) from an infant born with congenital scoliosis, who died at 9 days of age, was assessed. From the left to right assessments of the same segment by different techniques included a drawing of the vertebral deformation and tissue constituents of the vertebrae and discs, a specimen radiograph, histological preparation with staining by hematoxylin and eosin reproduced in *black* and *white*, and histological preparation with safranin O-fast green. In the drawing, the regions in *black*, cartilage; *black dots* on *white background,* endochondral bone of the vertebral bodies; *oblique lines*, intramembranous bone of the vertebral bodies; and *clear white areas*, nucleus pulposus of the intervertebral discs. In the safranin O-stained sections, *red*, cartilage of the vertebral bodies and (*lighter*) intervertebral discs and *green*, bone or fibrous tissue

vertebrae adjacent to the cartilage growth plates. Peripherally, the intramembranous woven bone forms the cortical walls. Marrow formation is normal. The T7-T8 and T8-T9 disc spaces have an almost normal-appearing nucleus pulposus and annulus fibrosus. The T9 vertebra is slightly asymmetric with the height of the body greater on the convex side. The hypertrophic chondrocyte sequence is intact through the entire 360° arc and there is no cortical membranous bone. The T9-T10 disc space is not as thick as normal, but the nucleus pulposus is present and centrally positioned. The peripheral laminae of the annulus fibrosus are not as well delineated as those above. The T10 vertebral body is smaller than expected and slightly irregular in shape with height greater toward the convex side. The hypertrophic chondrocyte sequence is present throughout the entire 360° arc but the body is eccentrically positioned in the cartilage mass closer to the concave side. The T10-T11 disc space is very thin; the nucleus pulposus is scanty and peripherally the annulus fibrosus is absent. Excessive cartilage persists in what should be the disc space. The nucleus pulposus is seen between T10 and the larger T11 body but absent adjacent to the

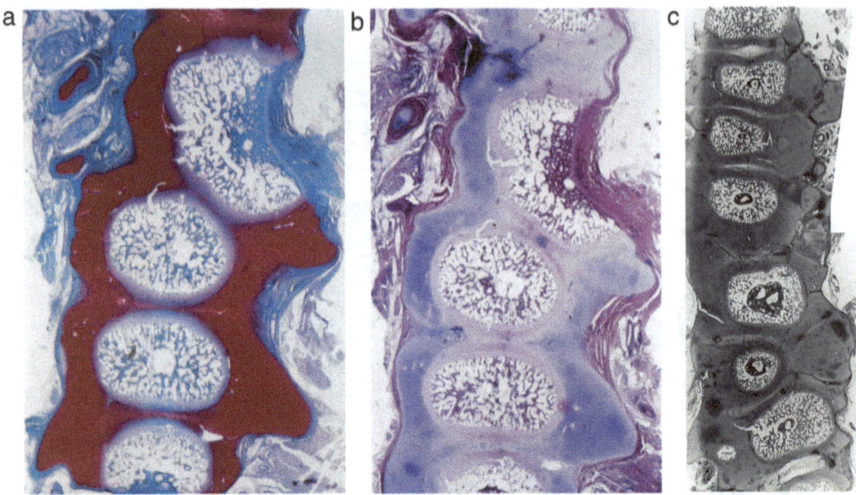

Fig. 3.16 Higher magnification views of the histological preparations of the congenital scoliosis of the spine in Fig. 3.15 are shown. The tissue was embedded in paraffin. (**a**) Safranin O-fast green preparation from L4 to S2 region. (**b**) Hematoxylin and eosin preparation from the same region shown in (**a**). (**c**) Hematoxylin and eosin preparations (montage, in continuity) from T7 to L4 [(**a**), (**b**), and (**c**) reprinted with permission from Shapiro F, Eyre DR. Congenital Scoliosis. A histopathologic study. Spine 1981; 6: 107-118. Lippincott-Raven Publishers]

smaller of the two T11 bodies and its ossification site. The T11 vertebra is composed of two separate and unequal ossification centers. Endochondral sequences with prominent hypertrophic chondrocytes are present in both and the endochondral bone is formed. The T11-T12 intervertebral disc is thicker than normal, particularly in relation to the smaller osseous body. The nucleus pulposus is most prominent between the larger of the two osseous areas and T12. The annulus fibrosus is lacking and the cartilage persists. The ossification area of the T12 vertebra is slightly eccentric, being positioned toward the concave side. The hypertrophic chondrocyte sequence is intact throughout a 360° arc and the endochondral bone is formed. At the central inferior aspect of this vertebral body, there is a small endochondral ossification area. The T12-Lumbar1 (T12-L1) interspace is markedly increased in thickness; the annulus fibrosus is not present peripherally and the cartilage tissue persists. A small nucleus pulposus is seen at the inferior margin of the disc space just above the L1 body. The L1 vertebra is not well aligned with T12, being present toward the convex side. It retains its generally oval shape although the height on the convex side is slightly greater than on the concave. Hypertrophic chondrocytes are present in a 360° arc. Centrally, there is a thick mass of histologically normal woven bone. Peripherally, the L1-L2 disc space on the concave side is narrower than expected but a nucleus pulposus is seen. Toward the convex side, just superior to the margin of L2, the nucleus is irregular, and an undifferentiated cartilage mass persists. Peripherally, the annulus fibrosus has not formed normally on the concave side and has not formed at all on the convex side.

The L2-L3 space is very thin and oblique and lacks both a normal nucleus pulposus and an annulus fibrosus. Only a trace of the nucleus pulposus is seen. Above L3 toward the convex side, a small nucleus pulposus is present. The L3 vertebra is irregularly shaped but the hypertrophic cell sequence is intact. It is not well aligned with the vertebrae above and is eccentrically positioned in the cartilage mass to the convex side. A smaller endochondral ossification center has begun to develop in the cartilage mass on the concave side of L3. The L3-L4 nucleus pulposus is much wider and thicker than normal. The annulus fibrosus is markedly abnormal as well, being more extensive and with annular fibers which are well organized and much more prominent (thicker) than usual. The L4 vertebra is malaligned toward the convex side of the cartilage mass. It is wedge shaped and peripherally the endochondral sequence has been replaced by the intramembranous bone. The nucleus pulposus between L4 and L5 is much more extensive than usual and is associated with a thicker annulus fibrosus at both concave and convex peripheries. The L5 and S1 vertebrae are developing as one osseous center. A unilateral bar of the intramembranous bone is present peripherally on the convex side. The two centers are in continuity and the endochondral sequence is intact throughout an approximate 260° arc. There is no nucleus pulposus between sacral 1 (S1) and S2. An annulus fibrosus has not formed on the concave side but is present on the convex side. The S2 vertebral body, S2-S3 interspace, and the S3 and S4 vertebrae are all normal.

Variation in Tissue Structure at the Same Level as Demonstrated in Serial Sections

Considerable differences in tissue structure at the same level are observed in the serial histologic sections, indicating that the histologic abnormalities are more variable than the two-dimensional impression obtained from a single radiograph indicates. The T7, T8, and T9 vertebrae and their intervertebral discs are constant in appearance throughout the 70 serial sections. Vertebrae T10 to L3, however, show the woven bone in the central area of the bodies on some of the serial sections, but as the sections pass more deeply inward, the size and frequency of occurrence of the woven bone mass progressively diminish, and in several of the sections, no bone is seen. Vertebrae T10, T11, and T12 are narrower than the thoracic vertebrae above in the coronal plane. In the serial sections, their size progressively and significantly diminishes, indicating a decreased size in the sagittal direction. In some sections the L1 and L2 bodies are bifid with no bone or cartilage centrally and two centers of endochondral ossification peripherally. In some sections, there are two ossification centers for L2 and L3; in most there is just one. The L4 vertebra remains constant throughout, as do the fused L5 and S1 and the S2, S3, and S4 vertebrae. There is no nucleus pulposus between S1 and S2, and in certain sections two endochondral sequences are sufficiently close that they will eventually merge into one bone mass similar to the two vertebrae discussed above. In certain sections the ossification center of T11 is malaligned but single, yet in other sections a second area of endochondral ossification to the convex side of the major areas appears. There is no normal-appearing intervertebral disc tissue between the smaller osseous center and the vertebrae above and below, and the normal-appearing cartilage persists. In a few sections, a small ossification center with normal endochondral bone is present just inferior to T12. In some sections, there is notochordal tissue in the same place where endochondral ossification forming bone tissue is present in other sections.

Chapter 4
Discussion

4.1 Normal Mouse Vertebral Development Compared to Pudgy Mouse Vertebral Development: General Considerations

Each of the radiographic, histologic, whole mount, and three-dimensional computerized reconstruction techniques throws specific light on the vertebral, intervertebral disc, and rib abnormalities in the development of the pudgy mouse. The end result demonstrates a failure of both normal formation and normal segmentation in the pudgy mouse recognized now as a recessive genetic disorder in which mutations in the *Delta-like 3 gene (Dll3)* have been defined [2]. The variable appearance of the vertebral and rib abnormalities from mouse to mouse, including in particular the variable appearance in affected littermates, indicates that the gene abnormality alone does not account solely for the deformities seen. Once the gene abnormality is expressed early in embryogenesis, secondary effects, presumably by mechanisms of epigenetics, appear to play a significant role in outlining the abnormal pattern. The pudgy mouse and other genetically triggered axial developmental abnormalities represent excellent models to help unravel pathogenetic mechanisms whereby gene abnormalities are translated into three-dimensional structural abnormalities.

The radiographs provide a two-dimensional picture and the whole mount preparations a three-dimensional appreciation of the gross structure of the vertebral and rib abnormalities. The histologic sections assess the cell and matrix components in a narrow depth since individual sections are only 5 μm thick, but serial sections, examined sequentially, provide a superb view of the cascade of adjacent changes, and the three-dimensional computerized reconstructions of the serial histologic sections allow the individual components of the axial structure to be seen in isolation or in relation to some or all of the other structures.

In the radiolucent regions between the osseous vertebral bodies, the tissue as indicated by histologic studies can be either intervertebral disc material or cartilage.

F. Shapiro, *Disordered Vertebral and Rib Morphology in Pudgy Mice*, Advances in Anatomy, Embryology and Cell Biology 221, DOI 10.1007/978-3-319-43151-2_4

In serial histologic section assessments and reconstructions, the patterns of bone tiisue, cartilage, and disc deformity are much more complex than appreciated on plain radiographs. The axial abnormalities lead to a variable mix of deformities including *hemivertebrae*, *wedge-shaped vertebrae*, *bifid vertebrae* (butterfly vertebrae or double hemivertebrae at the same level), and *fused vertebrae* (referring to continuous bone union between two adjacent vertebrae: *block vertebrae* if there is no interposition of the disc tissue at all between the bones and a *unilateral bony bar* if the bone continuity is only partial across the width of the bones). These abnormalities can be variable from segment to segment and at varying positions within the same segment (dorsal to ventral and/or cranial to caudal). Assessment of the embryogenesis of vertebral abnormalities is greatly aided by an understanding of cell, matrix, and tissue deposition patterns throughout the entire extent of the abnormal segments.

Histologic examination of the cartilage, bone, nucleus pulposus, and annulus fibrosus tissue shows the cells and immediately adjacent matrices in each of these tissues in isolation to be normal. The developmental problem of malformation occurs early in embryonic development leading to markedly abnormal shaping and/or positioning of the tissues. The primary gene or environmental (+/− gene) abnormality for the pudgy disorder is expressed within the unsegmented presomitic mesoderm (PSM) and then the segmented somites and sclerotomes at dorsal, ventral, or lateral positions leading to an abnormal positioning of cells responsible for axial development. It is now widely accepted that single gene mutations, discovered increasingly over the past two decades, initially trigger the malformation cascade. Previously (over several decades) it was regional tissue accumulations that had been shown as positive inductive influences in axial embryogenesis when present and negative influences when removed completely, repositioned, or made structurally abnormal. It was implicitly recognized by most researchers, however, that it was the (at that time unknown) molecular constituents of those regional structures that were the inducing agents.

It was previously considered, and still remains theoretically possible, that this cell pool (presomitic mesoderm/somites/sclerotomes) is fundamentally normal but is directed or forced secondarily to pattern cells involved in axial development by abnormalities concentrated primarily in adjacent parts of the developing embryo such as the notochord or neural tube which are known to play a specific inductive role in embryonic pattern formation. The notochord forms very early in embryonic development (before the somites) and serves as (i) a support structure along the dorsal midline for the developing embryo, (ii) a scaffold around which the undifferentiated mesenchymal cells of the sclerotome are positioned and subsequently differentiate to form the cartilage model of the developing vertebral bodies, and (iii) an active inducing agent in the differentiation of the surrounding tissues very early in development. It was previously demonstrated that notochordal abnormalities could act upon an otherwise normal collection of somite and/or sclerotome cells causing them to develop in an abnormal way, either on the basis of a chemical or positional abnormality. It was postulated that a nonlinear, absent, or molecularly deficient notochord could then induce molecular changes along the presomitic mesoderm/somite/sclerotome line with subsequent formation of cartilage vertebral bodies in abnormal positions and shapes. It is well known that neural tube/spinal cord abnormalities have a high association with congenital vertebral defects in humans but this region is normal in the pudgy mouse.

Vertebral development in the mouse has been the focus of much study for several decades with the classic works of Dawes [14] and Theiler [15] encompassing multiple histologic descriptions. Theiler also defined a wide range of axial developmental abnormalities of genetic origin in several mouse species in a classic monograph [16]. He felt these were of notochord, somite, or neural tube origin. We observed that, while each affected pudgy mouse had marked structural irregularities throughout the entire vertebral column including the tail and the ribs, the pattern of findings was different in each mouse. This confirms the original observation expressed by Grünberg [1] that "the general type of anomaly of the vertebral column is very uniform in all pudgies examined. But as the state of the axial skeleton can only be described as chaotic, no two pudgies agree in detail." Studies of affected and non-affected pudgy siblings sacrificed at the same time have extreme importance in terms of understanding and demonstrating the mechanisms of development of axial malformations. In each of the individual pudgy mice assessed, including those within eight sibling groups, the same pattern of structural differences involving vertebral and rib abnormalities was never seen. Since the gene abnormality would be the same throughout the pudgy group, including in particular mutants born in the same litter, it is important to interpret this finding that the gene mutation eventually translates into different structural abnormalities with markedly different patterns of vertebral and rib irregularities. We interpret these changes to indicate that structural abnormalities, while dependent initially on the specific defined gene abnormality, also develop on the basis of secondary inductive or epigenetic changes and even tertiary biophysical/biomechanical changes. These results indicate that components of biological structure, normal and abnormal, appear to be determined by epigenetic actions rather than being patterned completely as the result of an individual mutated gene expression.

The radiographic appearance shortly after birth does not fully predict eventual deformation. Where bony areas are present, definition is clear. In those regions, however, which are radiolucent, the actual tissue present as indicated by histologic studies can be either intervertebral disc material involving the nucleus pulposus and the annulus fibrosus or cartilage that has not yet ossified. There are significant developmental implications based on the amount and positioning of these two tissue groups between the already differentiated bone tissue regions. If the tissue between the developing bones is intervertebral disc, then the two bone centers will remain separate and an intervertebral space (however abnormal) will form. In those areas where there is no disc material but rather a continuous cartilage mass, conversion completely to bone tissue via the endochondral sequence allows for bridging of what should have been an intervertebral disc space by a continuous block vertebra or partial bony bar. Radiolucent tissues are differentiated in this study by histology but can be differentiated clinically early in postnatal development (and even at late fetal stages) by magnetic resonance imaging.

4.2 The Study of Vertebral/Axial Development Showing Histologic and Molecular Progression

4.2.1 Concepts and Mechanisms of Embryonic Vertebral Development

In the early nineteenth century, von Baer pointed out that embryologic development occurs by a process whereby undifferentiated tissue masses become progressively more differentiated and organized, *each stage being dependent on and channeled by the preceding stage* [17]. Early microscopic study then assessed the *deposition of tissues and cells* forming recognizable structures. Remak in 1855 described and illustrated chick embryo development, especially relating to the spinal cord and vertebral column, introducing the concept of vertebral *resegmentation* in a way still relevant (and debated by some) today [18]. Von Ebner described the *intervertebral fissure* in 1886 [19], a structure implicated in the resegmentation controversy since then, gradually being recognized as real, but still doubted by some as a tissue preparation artifact or simply a tissue interface as distinct from a true structure. Investigations into morphogenesis starting early in the twentieth century as experimental embryology took hold worked extensively on the concept of *biochemical gradients* as influencing embryonic tissue deposition patterns and, in a similar vein, tissue regions (e.g., notochord) serving to induce adjacent tissue differentiation. The *concept of induction* has also been important in explaining embryologic development, certain tissues augmenting existing potentials in adjacent less differentiated tissues to hasten the rate and ensure the patterns and uniformity of development. More recently, embryologic studies determined that genetic direction was based on recognition of *compartments within which development occurs* [20] and on *positional information* [21] as cells differentiated along lines based on their awareness of their positions. The development occurs as cells differentiate on the basis of their positions in time and space such that undifferentiated cells, for example, with skeletal development, become either chondroblasts or osteoblasts partly on the basis of their genetic constitution and partly on the basis of what cells immediately adjacent to them have done. Genetic directions for regions or areas (referred to as compartments) were postulated as mediated by genetic programing at the molecular level. A *clonal model* explaining vertebral column development was proposed by Moore and Mintz [22]. Based on an analysis of 35 genetically mosaic skeletons of four-parent allophenic mice comparing bone morphology with that of pure-strain controls, they concluded that there were at least four units of independent morphological variability and therefore a minimum of four developmental cell lineages in any vertebra. They felt either that the caudal and cranial sclerotomites are the clonal units capable of separate gene control or that the entire sclerotome component from each somite is the clonal unit. Since we demonstrate that each pudgy mouse has a different abnormal vertebral, intervertebral disc, and adjacent rib structure, it becomes both necessary and possible to explain these findings not only by more recently identified gene mutations but also by reintroducing previously marginalized concepts such as *Waddington's epigenetic pathway or epigenetic landscape* [23, 24] as well as current theories of *oscillatory mechanisms underlying segmental development (the segmental clock)* [25]. Waddington sought to determine how cells decide on which developmental trajectory to take since he felt that not all developmental events could be explained by gene actions alone. He defined "epigenetics" as that branch of biology assessing the causal interaction of genes and their products leading to the phenotype. He derived an illustration of a ball beginning passage down an incline with peaks (ridges) and valleys where the slightest deviation of the ball's downward (or downward and sideward) trajectory led to differing pathways that in the biological sense led to different cell fates (Fig. 4.1). The concept of the "epigenetic landscape" is derived from his illustration but theoretical biologists are now developing ways of quantifying Waddington's landscape. Bhattacharya et al. are working on a quantitative map of the epigenetic landscape using a mathematical numerical method developing deterministic rate equations governing gene regulatory circuits and then using stochastic simulations to assess cell fates with low-lying valleys representing stable cell states and higher ridges acting as barriers to transitions between stable states [26]. Wang et al. developed a theoretical framework using path-finding algorithms where the dynamics of developmental processes are controlled by a combination of the gradient and curl forces on the landscape [27]. Much of the study on epigenetics remains in the realm of molecular biology. Epigenetics assesses the final outcome of a locus or chromosome without changing the DNA. The three primary molecular controls of epigenetic phenomena involve DNA methylation characterizing chemical modification of chromatin, covalent modifications of histone proteins, and assessment of noncoding RNAs [28].

A deeper understanding of segmental development was sought first by theoretical considerations and more recently by molecular localization (expression/inhibition) demonstrations. The *clock and wavefront model* proposed by Cooke and Zeeman explains segmental structure formation as being temporally controlled by molecular oscillators allowing for repetitive waves of stimulation and inhibition along the midline axis [29]. The model involves an interacting "clock" and "wavefront"; the "clock" being a smooth cellular oscillator with cells throughout the embryo phase linked, while the "wavefront" is a front of rapid cell change moving down the embryo axis. Cells enter a phase of rapid alteration in locomotory/adhesive properties at progressively later times according to their position. Many recent studies have outlined the *molecular basis of the oscillator, referred to as the segmentation clock*, which acts in the presomitic mesoderm (PSM) to direct somite formation in a rhythmic fashion with specific interaction and control by Notch, Wnt, and fgf8 signaling [30–35]. Somite formation is strictly time dependent, varying in different species (e.g., each bilateral somite in the chick takes about 100 min to form). This signaling, setting up segmentation as somites develop, has been noted to commence as early as the PSM stage with the transcription factor *c-hairy 1* sweeping once across the PSM as each somite forms. The best characterized cyclic genes are involved in *Notch signaling* and include the Notch ligand

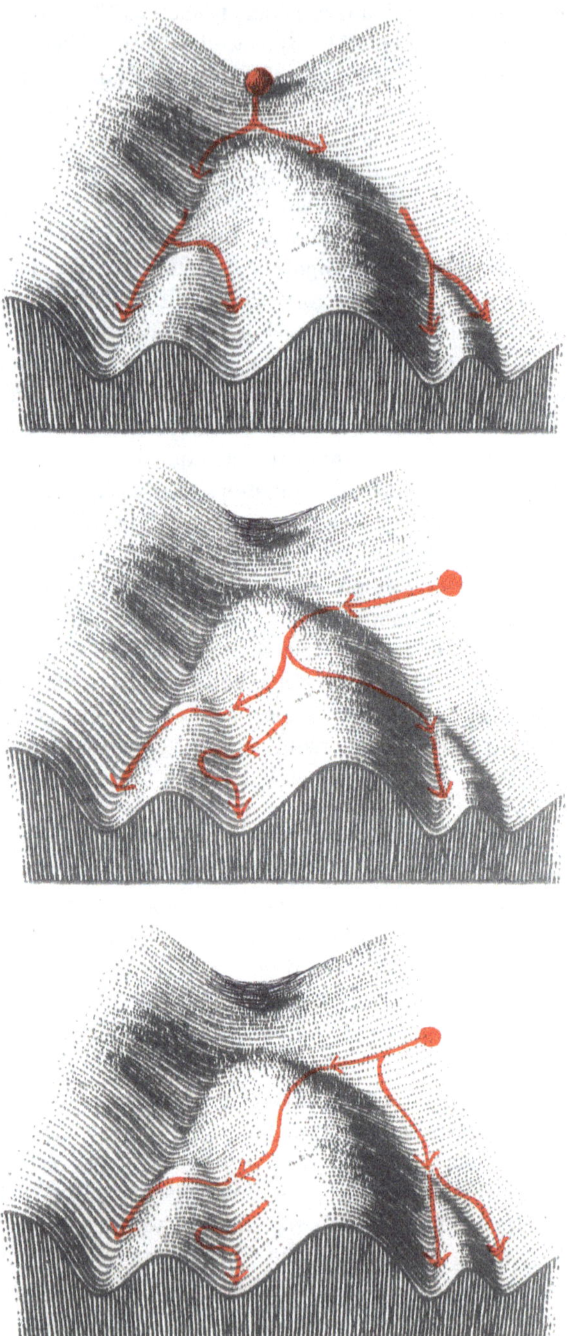

Fig. 4.1 Illustrations based on the drawing by Waddington of the epigenetic landscape help explain the concept of development along differing pathways (of cell and tissue differentiation) determined initially by a primary gene mutation, in effect sending the ball down the hill, but then by a varying series of minute changes at points of intersection along the pathway where one

delta C, transcription factors of the *hairy and enhancer of split (HES)* family genes, *the hairy and enhancer of split-related (HER) repressor genes Her1 and Her7*, and the glycosyl transferase *lunatic fringe (lfrg)* [36, 37]. Other genes involved in this activity are linked to the *Wnt signaling* pathway, while the growth factor *fgf8* has also been identified in cyclic activity [38]. Baker et al. developed a mathematical formulation of a new version of the "clock and wavefront" model based on the consideration that fibroblast growth factor *(FGF)* signaling controls somite boundary position and regulates segmentation clock control of spatiotemporal *Hox* gene activation. FGF8 expression in the presomitic mesoderm acts as the wavefront of cell determination as it moves along the body axis directing cells into somites. Pourquié has commented: "in all species examined thus far, Notch activation lies at the heart of the oscillator" [34]. As each somite forms, oscillating gene expression is repeated; expression and inhibition occur in a constrained temporal and spatial way. Much of this theoretical framework has been validated by gene recognition wherein specific genes are expressed in certain temporal and spatial patterns, leading to their implication as causative agents. Equally importantly, gene mutations help to both define functional implications and explain the development of malformations.

4.2.2 Histogenesis of Vertebral Pattern Development

In vertebral pattern development in several species, the anteroposterior (cranial-caudal) body axis is generated with formation of similar repeated structures or segments [40–42] (Fig. 4.2). Embryonic development progresses from the earliest primitive streak to bilateral non-segmented paraxial mesoderm (presomitic mesoderm, PSM in mice; segmental plate mesoderm, in chick), to segmented spherical somites (bilateral, paraxial structures which produce skeletal muscles, vertebrae, and dermis) [43–50], and also accompanied by regional formation of the notochord, neural plate/neural fold/neural tube/spinal cord, and ganglia/spinal nerves [51–63]. Somitogenesis is followed by subdivision (where cellular/molecular regions or compartments control specific cell fates): the dorsal half of the somite into

Fig. 4.1 (continued) direction or another must be taken. The *red ball* at the *upper* or *right* sides of each "landscape" represents an undifferentiated cell in the course of its downward passage down valleys (*troughs*) and over peaks. The points of divergence lead cell reactions in initially slightly different directions whose differences are magnified as passage along one pathway or another continues. The landscape can easily be considered as a pathway traveled down a ski hill. The initial mutation (e.g., in the delta like three genes or the pudgy costo-vertebral deformation) leads to the disruption of the normal timing and synthesis pattern of segmentation, but the more subtle changes which cause a specific difference between each pudgy mouse are determined by the minute epigenetic changes demonstrated by the landscape theory of Waddington. Each of the three images shows how minute pathway changes at various points of the downward course can lead to progressively more different outcomes

dermomyotome and the ventral half into the sclerotome. The dermomyotome is the formation source of the dermis of the back and muscle cells of several regions. The sclerotome is then compartmentalized into dorsal and ventral sclerotome medially (which gives rise to the vertebral body and intervertebral disc), the lateral sclerotome (which gives rise to neural arches, ribs, and vertebral pedicles), as well as the central sclerotome that has a cranial (rostral, anterior) half and a caudal (posterior) half. The intersegmental fissure of von Ebner separates the anterior/posterior segments of the sclerotome but, in the opinion of some, *does not* extend further medially to the axial area where the vertebral bodies will form. Others feel that it does, helping to delineate the collections of cells lightly versus densely packed and actively involved with resegmentation. Resegmentation then occurs as vertebral bodies and intervertebral discs form [see below], while laterally rib formation occurs (also segmentally determined), and mesenchymal tissues differentiate to allow for cartilage and bone tissue formation (Fig. 4.2a–d). Neural arches are derived from the caudal portion of one somite and the vertebral body from parts of two adjacent somites. Segmentation develops sequentially in a craniocaudal direction with the caudal (tail/posterior) somites the youngest and the cranial (head/anterior) the oldest [64–66]. Further regarding each somite, the medial half is derived from the lateral part of Hensen's node, while the lateral half is derived from the primitive streak. Segmentation also encompasses formation of such structures as the dorsal root ganglia, spinal nerves, muscles, and blood vessels. The various regions are associated with specific molecular localizations that then define the structural changes.

Detailed reference to vertebral development has been made in descriptions of chick embryology [64–66]. Remak described chick embryo development in a classic work in 1855 [18] in which he outlined the vertebral body development theory of resegmentation. Somitogenesis in the paraxial mesoderm has been well described by Tam and Trainor [67]. Articles from the classic descriptive era of embryology to the current molecular era have repeatedly assessed axial development in relationship to the concept of resegmentation with strong opinions voiced pro and con regarding its accuracy [18, 19, 47, 68, 69]. Remak considered that the early embryological collections of midline cells, which he referred to as protovertebrae, were the first manifestations of what would be the adult vertebral bodies. That term was replaced with the descriptive term "somite" in the late 1800s. Remak noted that the protovertebrae (somites) did not correlate exactly in position with the final vertebrae and that a resegmentation occurred during development. Most observers today accept that the boundaries of the vertebrae are shifted one-half segment in relation to the corresponding somites. The somite demonstrated a cranial (anterior) collection of lightly concentrated cells and a caudal (posterior) collection of more densely concentrated cells. As development proceeded, some of the densely packed cells cranially formed the intervertebral disc, while the remaining dense cells merged with the next caudal collection of lightly packed cells to form a single vertebral body. As expressed by Christ and Wilting "a vertebra is formed by the combination of the caudal half of one bilateral pair of somites with the cranial half of the immediately caudal pair of somites"

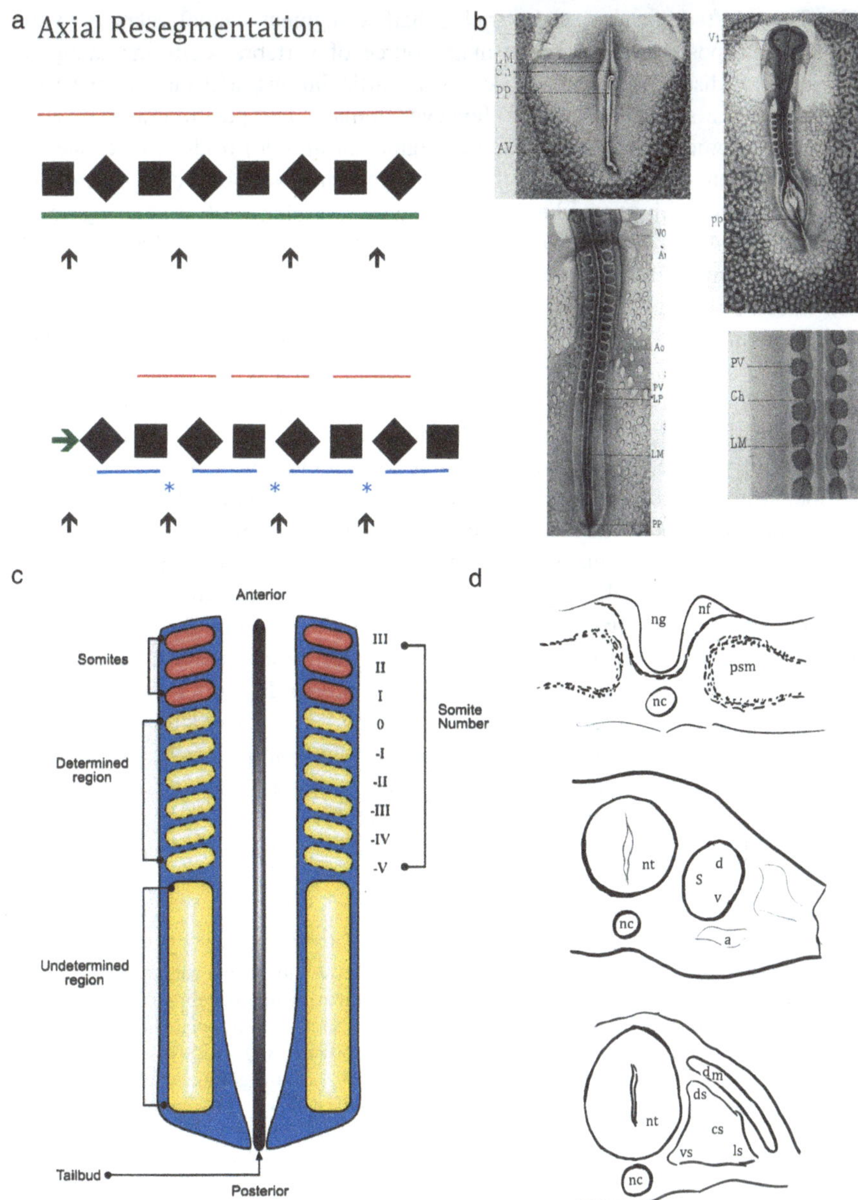

Fig. 4.2 An overview of somitogenesis and progressive embryonic development is presented in this figure. (**a**) Somitogenesis. *Code*: (*filled square*) anterior sclerotome, (*filled diamond*) posterior sclerotome, (*solid red line*) myotome (*top*), muscle (*bottom*), (*asterisk*) intervertebral disc, *upper arrow* represents corresponding regions in *top* (somite/sclerotome) and *bottom* (vertebral body) stages of development. *Arrows* (along the same axis) at the top delineate anterior from posterior halves of each sclerotome and at the bottom the intervertebral discs between vertebral bodies [with resegmentation each vertebral body forms by recombination of a posterior (caudal) sclerotome and the next adjacent anterior (cranial) sclerotome]. This formation from adjacent segments causes a half-segmental shift in the alignment of sclerotomal segments compared with the more mature vertebral bodies. The myotomes retain their original segmental positions when

[43]. This shifted the eventual vertebra half a segment caudal relative to the somites. The posterior half is the main source of vertebral cells including the intervertebral discs, transverse processes, rib articulations, and most of the facet joints [47]. The somatic myotome, however, remained in position, and after the vertebral resegmentation, the myotome (segmental muscle) bridged or connected the two adjacent vertebrae across the now formed intervertebral disc. The matter was further defined by von Ebner's demonstration (in 1885) of an intersegmental fissure lying inferior to each caudal dense collection [19]. Opinion has gone back and forth regarding the accuracy of Remak's observation and whether von Ebner's fissure was a true anatomic structure or just a tissue-processing artifact. Presently, the resegmentation theory is accepted by most with chick-quail chimera transplant studies [48] being a prime reason, but detailed negation of the theory was being offered only a few decades ago. Baur [70], Verbout [69], Dalgleish [71], and Theiler [16] felt that resegmentation did not occur. All opinions and findings on this issue of resegmentation with vertebral development were well reviewed by Verbout who concluded that "a development of the axial skeleton without resegmentation is just as conceivable in functional terms as one in which resegmentation occurs" and that the literature offers no convincing support for any version of the "Neugliederung" (resegmentation) concept which he felt was "in all probability invalid" [69]. Those accepting that view felt that the vertebrae, and other associated segmented structures (ribs, ganglia), "arise directly in their definitive places and levels." Dalgleish, in a study of mouse development, felt that the cleft was simply an interface between the cranial and caudal halves of the sclerotome and that "the vertebral bodies developed from primarily unsegmented tissue surrounding the notochord and lying within the axial mesenchyme…tissue…variously termed perichordal tube or perichordal sheath…." [71]. Currently, however, Christ et al. [45, 47] and Fleming et al. [42]

Fig. 4.2 (continued) they become mature muscles (i.e., they do not resegment); *solid green line*, notochord in early embryo; *green arrow*, remnant of notochord within vertebral column (nucleus pulposus at intervertebral discs); *blue lines* (*bottom image*), united sclerotomal segments forming individual vertebral bodies to complete resegmentation process. (**b**) Images reproduced from Mathias Duval, Atlas d'Embryologie (Paris, G. Masson, 1889), show good understanding in that era of the progressive nature of axial development. The illustrations are from a study of chick embryo development. The primitive streak is shown at the upper left, progressive somite (referred to as a "prevertebra") development at the upper right and lower left, and then an enlarged illustration of development with somite differentiation as well as the prominence of the notochordal and neural tissue in midline development. *Code*: *pp* ligne primitive, primitive steak; *Ch* la corde dorsal, notochord; *PV* prévertèbre, prevertebra (our somite), *LP* lame prévertèbrale, preveretbral plate, *LM* lame medullaire (developing neural tube). (**c**) This concept was outlined recently in relation to molecular findings [reprinted with permission from Eckalbar et al. Scoliosis and segmentation defects of the vertebrae. WIREs Dev Biol. 2012: 1: 401-23, Wiley, Inc.]. (**d**) A series of drawings highlight vertebrate axial development from earlier (*top*) to later (*bottom*) time frames. *Code*: *ng* neural groove, *nf* neural fold, *nc* notochord, *psm* presomitic mesoderm, *nt* neural tube' *a* aorta, *S* somite, *d* dorsal, *v* ventral, *dm* dermomyotome, *ds* dorsal sclerotome, *cs* central sclerotome, *ls* lateral sclerotome, and *vs* ventral sclerotome

strongly support the validity of the resegmentation theory in which there is a realignment of segmentation between the somite stage and the vertebral stage expressed again as showing the sclerotomes rearranging (resegmenting) by fusion of the caudal (dense) half of one sclerotome with the cranial (light) half of the following adjacent sclerotome.

Vertebral development has been studied in urodeles (salamanders), chick, birds, mouse, sheep, and human. Although there are significant differences, the overall developmental pattern is similar. The chick embryo is relatively easy to study and is amenable to experimental surgical manipulation in vivo, while the mouse is favored for greater similarity to the human structurally and for gene and molecular studies. Vertebral developmental patterns and histogenesis have been well defined among hundreds of articles, some of the most relevant for this presentation including birds (gulls, ostriches) by Piiper [72]; chick by Romanoff [65], Shapiro [13], Christ and Ordahl [44], and Christ et al. [45]; mouse by Dawes [14] and Dalgleish [71]; sheep by Verbout [41]; human by Bardeen [73], Ehrenhaft [74], Wyburn [75], Sensenig [76], Bick and Copel [77], Tondury [78, 79], and O'Rahilly and Meyer [80]; chick and human by Christ and Wilting [43]; chick primarily (with reference to the work in other species) by Hall [51]; and multiple species by Baur [70]. The development of the intervertebral disc in the human has been well described by Keyes and Compere [81], Walmsley [82], and Roberts et al. [83].

Arcualia Theory A theory of vertebral formation from four pairs of tissue collections called arcualia was proposed by Gadow and Abbot [84] and Gadow [85] based on primary assessments from fish vertebrae. The description was prominent for several decades as multiple observers attempted to fit all vertebrate development into that framework [86]. Piiper wrote an extensive, detailed work on vertebral development in the gull embryo (which is similar to the chick) from 2 to 10 days [72]. His descriptions and illustrations of development were superb, but the work was "frighteningly complex" according to Williams [86] because his analysis of vertebral development used the arcualia theory terminology onto which his own categorizations were added. Vertebral development in fish was defined as being representative of essentially all vertebral development with the vertebral column composed by repetition of four pairs of tissue collections referred to as arcualia. The pairs of arcualia were termed basidorsalia, basiventralia, interdorsalia, and interventralia. Williams [86] and Verbout [69] critically reviewed Gadow's arcualia theory, finding it completely inappropriate as a model for vertebrate development, and it is referred to today only in a historical sense owing to the prominence it once held.

4.2.3 Timing in Embryonic Axial Development

Jurand describes mouse notochord development being underway at 8 days as prenotochordal cells form from the cranial end of the primitive streak in the

Hensen's node area [61]. The development proceeds in a cranial to caudal direction. The notochord is not derived from the presomitic mesoderm but develops separately at the same time. Theiler also indicates that the critical period for somite formation in the mouse is between 8 and 11 days [16]. Since the somites develop in a cranial to caudal sequence, cervical development is occurring at 8.5 days, thoracic 9 days, and lumbosacral 10 days. In the human this critical period extends from 23 to 26 days. O'Rahilly and Meyer detailed the timing and sequence of events in the embryonic development of the human vertebral column from 16 days to 57 days [80]. The notochordal process appeared at 16 days and progressively developed as the notochordal canal (18 days), the notochordal plate (20 days), with the notochord developing to the stage of first appearing as a solid cord and extending to the caudal region at 22 days. Regarding the somites: sclerotomic cells were migrating from the somites at 22 days; occipital, cervical, and thoracic somites were distinguishable (in a 20 somite embryo) at 24 days; additional lumbar somites were distinguishable (in a 24 somite embryo) at 26 days; additional sacral somites were distinguishable (in a 28–29 somite embryo); and some additional coccygeal somites were present (in 35 somite embryos) at 28 days. At 33 days, 38 somite embryos had developed. Ganglia were not described as appearing until 28 days (more than 20 pairs) with the adult complement of ganglia present at 32 days. Intersclerotomic fissures were seen in most embryos, and cranial and caudal sclerotomites were distinguishable at 28 days with the caudal sclerotomites becoming denser (more condensed) than the cranial at 32 days. The rib anlagen first appeared at 28 days.

Hautier et al. [87] have studied the ossification sequences of the neural arches and vertebral bodies in normal mice (wild-type CD-1). The sequence of ossification in the human fetus (neural arch and vertebral bodies) reveals a similar pattern in the study by Bagnall et al. [88]. O'Rahilly et al. studied the state of human vertebral column development at the end of the embryonic and early fetal periods [89]. In both humans and mice, ossification of the vertebrae does not follow a cephalocaudal (cranial-caudal) pathway, and the ossification centers for the neural arches and the vertebral bodies (centra) develop independently. In the human, Bagnall et al. demonstrated that ossification occurs first in the lower thoracic/upper lumbar vertebrae and then spreads in cranial and caudal directions (cranial more rapidly). The T11, T12, and L1 vertebral bodies ossify first and at the same time. For the neural arches, the lower cervical/upper thoracic arches ossify first followed by the upper cervical and then the lower thoracic/upper lumbar after which the ossification spreads to the intervening regions [88]. In mice, Hautier et al. observed similar, but not exact, patterns of ossification using high-resolution X-ray microtomography (micro-CT) [87]. As in the human, a strict cephalocaudal ossification pattern is not seen. In mice, the vertebral bodies (centra) always ossify after the neural arches. For the vertebral bodies (centra), the lower thoracic vertebrae ossify first (T10 to T13) after which ossification spreads cranially and caudally (although faster caudally in the mice). Ossification of the neural arches proceeds from at least two regions, first in the upper cervicals (C1 to C6) and second in the lower thoracics with ossification then spreading between these two initial regions to meet in the middle of the rib cage around the sixth rib.

4.2.4 Cell and Tissue Patterns in Vertebral Development: The Notochord and Other Regions Inducing Adjacent Embryonic Tissue Development

In vertebral development in multiple species, early notochord development is seen accompanied by a prominent differentiated neural tube dorsal to it and then sequentially by formation of presomitic mesoderm (PSM), somites, sclerotomes, and cellular differentiation into the mesenchymal, cartilaginous, and osseous framework of the vertebral bodies and neural arches surrounding both the notochord and the neural tube/spinal cord, respectively. Much study over several decades has attempted to determine whether the close apposition of these tissue accumulations (notochord and neural tube/spinal cord, cartilaginous/bony vertebral bodies and posterior arches) in time (simultaneous and sequentially) and space pointed to an inducing role of one tissue group on another or to a simple coexistence with separate control mechanisms. It is widely accepted that the notochord enhances vertebral body segmentation and cartilage formation, while the neural tube/spinal cord underlies vertebral arch formation. Abundant evidence exists of the developmental interrelationships between the neural tube, notochord, and surrounding mesenchyme [51, 52, 56]. The normal development of the neural arches is dependent on the neural tube. Specifically, the central nervous system, including the ventral spinal cord and the spinal ganglia, has been considered to play a key role in inducing formation of the posterior bony elements (the neural arch). Evidence has also been presented that the pattern of cell differentiation along the dorsoventral axis of the chick neural tube is regulated by signals derived from the notochord and floor plate [57]. Simply stated, the notochord is felt to control sclerotome development.

The notochord has two broad functions in embryologic development: the first as an inducer for adjacent cells directing them to synthesize certain structures and the second in the structural/mechanical realm as an elongation support for the young embryo and as a scaffold around which vertebral cartilage differentiation occurs. The notochord helps direct craniocaudal organization along the embryonic axis and also serves as a primary organizer for multiple organs other than the spinal cord and vertebrae. The inducer role is most prominent in the early days of embryogenesis with the more passive mechanical role then predominating. Among the signals from the notochord, hedgehog proteins play important roles in embryogenesis. The notochord plays a role in coordinating dorsal-ventral development and in determining cell fates in somitogenesis. Specific to axial development, the notochord has been considered by many researchers to induce and control the appropriate formation of the vertebral bodies, being essential for morphogenesis (cartilage segmentation) but not for cartilage differentiation [51, 52]. These considerations arose from and were supported by experimental embryo studies. Surgical removal of embryologic structures helped determine their inductive effects. When the notochord of the developing chick was removed surgically, vertebral body development occurred but was markedly abnormal [51, 52, 56]. Others, however, doubted the significance of the notochord in positively affecting adjacent vertebral body formation based on experimental findings in other species [53, 90]. The notochord plays a key role in establishing the dorsoventral polarity of the neural tube. It induces the ventral neuroepithelium to differentiate into the floor plate that then induces differentiation of motor neurons in the ventral horns [62]. Surgical removal of the notochord reveals that the vertebral bodies form but are unsegmented while the neural arches are normal. Strudel removed the neural tube from early embryos and found that eventually no neural arches formed. Removal of the notochord resulted in the absence of the vertebral body but had no effect on the development of the neural arches or ribs.

4.2.5 Rib, Ganglia, and Intervertebral Disc Development in Normal and Pudgy Mouse Specimens

4.2.5.1 Normal Rib Development Compared to Pudgy Rib Development

Each rib has two ends with a thin flat curvilinear body or shaft between. The ends are referred to as *dorsal/vertebral/proximal* adjacent to the vertebra and *ventral/sternal/distal* some attaching to the costal cartilage and then onto the sternum and some being free or unattached ventrally. The proximal end has a head, neck, and tubercle. The head has an articular surface with two facets articulating with the bodies of two adjacent vertebrae spanning the intervertebral disc, while the neck and tubercle articulate with the transverse process of the corresponding vertebra. The distal end articulates with the costal cartilage, a bar of hyaline cartilage that prolongs each rib, before articulation with the sternum or, below the sternum, it lies unattached. The head, neck, and tubercle are referred to as the proximal part of the rib and the costal shaft as the distal part with further subdivision into vertebral and sternal segments. Efforts to define the origin of the ribs continue, with some differing conclusions reached. Christ et al. showed that the entire ribs (and intercostal muscles) are derived from the somites, but the question remained from which compartment of the somites the ribs were formed [91]. This could involve either the ventral sclerotome or the dorsal dermomyotome; the general view was that the ribs derived from the sclerotome (Christ and Wilting) [43] but Kato and Aoyama [92] felt that the distal parts of the ribs originated from the dermomyotome. Huang et al. carried out quail-chick chimera experiments and concluded that dermomyotome cells did not participate in rib formation; cells of the sclerotome formed the axial skeleton and all parts of the ribs; and while ablation of the ectoderm over the region, which affects dermomyotome development, led to rib malformations, it was the disturbed interactions between dermomyotome and sclerotome that were responsible for the malformations (rather than either acting alone) [93]. They concluded that the ribs derived from the sclerotome. Moreover, resegmentation of the vertebral bodies also applied to rib formation with the chimeras showing the tissue along the shafts originating from two adjacent somites. Aoyama et al. continued their investigations transplanting and separating developmental segments with aluminum foil in chick embryos [94]. They defined three compartments for each rib that developed under different controls: proximal rib depending on ventral-axial tissues, vertebro-distal rib depending on the surface ectoderm, and sterno-distal rib depending on the lateral plate mesoderm (which in turn depends on the surface ectoderm). The proximal part was thus dependent on the notochord and ventral neural tube, while the distal rib depended on the surface ectoderm. They left indeterminate whether this was partly dermomyotome dependent and interpreted their findings in relation to a new classification of somite derivatives defined by Burke and Nowicki as primaxial (vertebral ribs) and abaxial (sternal ribs) [95]. Primaxial refers to somite cells that develop in an exclusively somitic environment and abaxial to somite cells that develop into lateral plate mesoderm derivatives. Different patterns of ribs might be dependent on region-specific or species-specific morphogenesis. Evans' investigations supported the belief that the sclerotome compartment of all thoracic somites contributes cells to both proximal and distal parts of the ribs and that resegmentation was also involved with rib, as well as the vertebral body, formation [96]. The intercostal muscles arise from the ventrolateral dermomyotome.

4.2.5.2 Gene Components of Rib Development: Hox Role in Patterning of the Axial Skeleton

Hox genes play a major role in patterning the vertebral axial skeleton [97–100]. Wellik points out that the only component of axial structure not dramatically altered with loss of *hox* function is the ability to form ribs. The thoracic ribs are the only axial phenotype not clearly shown to be *hox* dependent. There is no evidence that *hox* genes influence rib formation. The control of rib development appears to be regulated through *Myf5*/*Myf6* expression [99]. *Pax-1* is responsible for proximal rib development [93, 94]. The sclerotome expresses *Pax-1* while the dermomyotome expresses *Pax-3*. There appears to be good recognition that *Pax-1* is responsible at least for proximal rib development. Regarding the distal rib, Aoyama et al. felt there was some evidence that the *Myf5* knockout mouse lacked distal ribs. *Pax-3*-deficient mice also exhibit severe distal rib defects [96].

4.2.5.3 Ganglia Development

An additional observation in this study, which was also alluded to by Grünberg, is the presence of the spinal ganglia in the pudgy mouse as a uniform continuous band of the tissue on either side of the neural tube rather than being segmented into individual oval collections [1]. As the ganglia develop temporally in advance of cartilage vertebral body differentiation, the possibility exists that abnormalities in these ganglia might predispose to disordered development of the vertebral anlage.

Fused spinal ganglia were noted in amputated mouse (am/am) embryos that corresponded with abnormal spacing of sclerotome condensations and late vertebral fusions [16]. Others however have felt that ganglia do not govern sclerotome differentiation [40]. Since our initial observations in this study only begin late in the embryonic period close to birth, the early developmental patterns were not assessed. In our study, however, two observations are relevant. (i) Although the pudgy mouse ganglia appear continuous in the embryo at low-power light microscopic observation, closer examination at high-power magnification always shows an intact, thin fibrous septum between adjacent ganglia. (ii) Normal non-affected mice initially show similar appearing developing ganglionic tissue masses with apparent continuity but at greater magnification thin septa become apparent.

4.2.5.4 Intervertebral Disc and Vertebral End Plate Development

Vertebral body growth plates on superior and inferior (cranial and caudal) surfaces underlie longitudinal growth [77, 101]. The mouse vertebral bodies grow longitudinally by the endochondral mechanism from these true epiphyseal growth plates at upper and lower surfaces. This mechanism is used in the human, mouse, and other vertebrates, but the tissue composition encompassing the intervertebral disc and the adjacent vertebral body differs in various species. In the human the cartilage growth plate merges with the immediately adjacent intervertebral disc fibrocartilage with no epiphyseal secondary ossification center or calcified epiphyseal cartilage intervening. At the opposite end of the spectrum, rabbits and larger animals such as cows and goats have a well-developed secondary ossification center that forms between the vertebral body surface growth plate (physis) and the intervertebral disc fibrocartilaginous region. In the mouse, our postnatal radiographs show a linear radiolucency at upper and lower ends of the developing vertebrae with a thin radio-dense linear collection between the growth plate and the adjacent disc. Histologic sections, however, show no actual bone in these regions, leading to the interpretation that a portion of the cartilage calcifies but does not transform or develop into bone tissue. Calcium deposits were not seen in our histological sections since the tissues were decalcified in association with fixation. The mouse vertebral growth plate is thus similar to the human but markedly different from the rabbit and cow. These characterizations of intervertebral discs and vertebral cartilage end plates have recently been outlined as well by Dahia et al. [102] and Zhang et al. [103] and need to be considered in investigational studies. We note that, in the normal mice, this thin radio-dense line is horizontal adjacent to the normal intervertebral discs, but in affected pudgy vertebrae, the line curves in relation to the misshapen vertebral body, an observation made in particular with hemivertebrae, wedged vertebrae, and the double hemivertebrae at the same level (called by some butterfly vertebrae).

4.2.6 *Vertebral Development in the Chick Embryo*

Vertebral development in the chick embryo from day 3 to day 19 has been described and correlated with the Hamburger-Hamilton stages [104] that are based completely on external characteristics [13]. Whole mount preparations (Fig. 3.13a–c) provide excellent three-dimensional histochemical views of vertebral development in situ, which correlate well with histologic studies (Fig. 3.13d–k). Paraffin-embedded sections stain well with safranin O-fast green that is particularly sensitive to early cartilage differentiation, which stains bright red, and early bone formation that stains deep green. Plastic-embedded sections, however, define cellular detail at the light microscopic level in a much clearer fashion. A detailed histologic outline of the stages of vertebral development is an important reference point for molecular studies and embryologic manipulations (Fig. 3.13k). There is a craniocaudal temporal gradient of development within the axial as well as the appendicular skeleton and not all vertebrae are at the same stage of development at one time. This gradient refers to the shaping of the vertebrae as well as to the onset of chondrification and ossification with cervical development maturing slightly in advance of thoracic and the sacrococcygeal area maturing last. This craniocaudal gradient is demonstrated in the whole mount embryo at 14 days where vertebral body bone formation, shown in red, is extensive in the upper cervical vertebrae, moderate in the lower cervical and thoracic vertebrae, and absent in the sacrococcygeal vertebrae, which stain blue because of the persisting cartilage. [Ossification in chick embryo vertebrae differs from patterns in human and mice vertebrae with a relatively direct developmental sequence beginning at upper cervical vertebrae and moving sequentially to thoracic, lumbar, and sacral-coccygeal regions.] The histologic and autoradiographic descriptions were concentrated in the thoracic, lower cervical, and upper lumbar vertebrae, adjacent regions that develop "in essentially the same manner... and almost at the same time" [65]. The development has been well summarized [65, 66]. Histologic sections show the neural tube to be completely formed and the notochord prominent by 3 days. The notochord then becomes relatively diminished with each day. By 17 days only small notochordal remnants remain within the vertebral bodies. Notochordal cell uptake of tritiated thymidine is intense at 3 days and remains so at 5 days but by 7 days is dramatically diminished. No uptake is demonstrated in notochordal cells from 9 days onward, attesting to the role of the notochord only in the early phases of vertebral development. The notochord plays an important passive mechanical role in the early stages, serving as a structural support from somite 24 onward and as a scaffold around which the cartilaginous vertebral bodies are formed. Its early development from stages 5 to 31 has been described by light microscopy as well as by scanning and transmission electron microscopy [57, 61, 105]. The uptake of ^{3}H-thymidine in the developing vertebral body, neural arch, and spinous process is continuous throughout the period of study and occurs not only in the surrounding perichondrium but throughout the cartilage tissue as well, except in the hypertrophic chondrocytes. This pattern attests to the occurrence of interstitial as well as appositional cartilage mass increase. The development of the vertebral body with respect to both cartilage differentiation and bone deposition precedes that of the lateral neural arch and dorsal spinous process elements. Cartilage cell hypertrophy begins in the vertebral body of the lower cervical, thoracic, and upper lumbar vertebrae at 8 days in the perinotochordal region. The hypertrophic changes then proceed laterally and dorsally toward the neural arches. Separate centers of cartilage hypertrophy develop on either side laterally within the neural arches, and a separate single center develops even later in the dorsal spinous process. Although bone formation subsequently occurs, the mechanism of endochondral ossification differs somewhat from that seen in mammals. In chick vertebrae, cartilage cells undergo hypertrophy but the cartilage matrix is removed extensively by advancing fronts of round cells and chondroclasts. Bone deposition still occurs on occasion in developing vertebrae on cartilage surfaces, but relatively large cartilage cell and matrix accumulations are covered, rather than the much narrower single calcified cartilage cores on which bone tissue deposited in mammalian bones. This extensive cartilage matrix resorption also characterizes chick long bone development. Williams described normal chick cervical vertebral development in the embryo to 7 days, indicating that after that time only differential growth of structures already present occurred [90]. Much of his work involved experimental transplantation of vertebral rudiments as a method of assessing the effects of various regions on development.

4.3 The Pudgy Mouse

4.3.1 Appearance

The pudgy mouse is smaller than its noninvolved sibling with the shortness due to malformation involving the axial region spine and thorax where the structural abnormalities are seen. The tail is also shorter due to the vertebral involvement and from birth onward is also twisted upon itself. The appendicular bones of upper and lower extremities are of normal shape including the joints. The shoulder and hip joints are normal even though they are quite close to markedly abnormal cervicothoracic and lumbosacral vertebrae, respectively. A few of the pudgy mice are somewhat frail and die early but most survive and move about actively. A mouse with the mutation is evident even in the late embryonic phase and is clearly recognizable at birth with shortened stature (Fig. 3.1). This includes shortness of the tail in the embryo although the prenatal tail tends to be straight.

4.3.2 Gene Mutations

The pudgy mouse mutation was linked to chromosome 7, and shortly afterward, mutations were detected in the Delta-like 3 gene *(Dll3)* of the Notch family [2]. In the human, Turnpenny et al. [106] mapped the gene for autosomal recessive spondylocostal dysostosis to chromosome 19q3, and shortly thereafter the pudgy gene *Dll3* was identified as the causative mutation in some human cases of spondylocostal dysplasia (SCD) [107]. With histological and molecular markers in pudgy mice, the pu/pu mutation was seen to disrupt the formation of morphological borders in early somite formation and rostro-caudal compartment boundaries in somites [108]. Kusumi et al. indicated that *Dll3* was expressed in presomitic mesoderm and somites as well as in the neural tube [109]. In the pudgy embryos, it was found that the morphological borders between somites and compartment boundaries were severely disrupted. The somites were observed to be irregular in size and shape at later stages. Spinal nerves and ganglia were also irregularly formed and unevenly spaced, but the notochord appeared "normal and unbroken" including assessment by sonic hedgehog (Shh) expression. Studies of the pudgy mice by Dunwoodie et al. also found abnormal dorsal root ganglia which were irregular in size and shape and fused [108]. In the study by Dunwoodie et al., loss-of-function mutations in the mouse Delta-like 3 gene (*Dll3*) were generated following gene targeting that led to the skeletal malformations [108]. Axial skeletal disorganization involved all vertebrae from cervical 1 to the tail and adjacent ribs.

The *Mesp2* (mesoderm posterior 2) gene is expressed in the mouse rostral presomitic mesoderm but is then downregulated immediately after formation of the segmented somites [110]. *Mesp2*-deficient mice were then generated by gene targeting and mice with abnormal fused vertebrae and fused dorsal root ganglia with impaired sclerotomal polarity developed. The earliest defect noted was lack of segmented somites. It was concluded that *Mesp2* also played a major role in regulating signaling systems mediated by notch-delta and FGF that are essential for segmentation [111]. Morimoto et al. did further studies showing that the *Mesp2* transcription factor participated in the "clock and wavefront" model for somite formation by suppressing Notch activity through induction of the lunatic fringe (*Lfng*) gene [112]. The activity in the presomitic mesoderm helped explain that the oscillation of Notch activity is arrested and translated in the wavefront by *Mesp2*. Mutated *MESP2* was then found to cause human spondylocostal dysplasia in several cases with the radiologic criteria but with no mutations in the *Dll3* gene [113, 114]. It appeared to be the major cause of this skeletal dysplasia in families of Puerto Rican origin [114]. *Lunatic fringe* was also found to function as a boundary-specific signaling molecule needed to localize activity by the protein Notch between individual somites. When mutants were generated, there was failure to form boundaries between individual somites and the axial skeleton was highly disorganized [115]. Mutation of the *LUNATIC FRINGE* gene is also identified in some human cases of spondylocostal dysplasia [116]. There are thus three gene mutations tied to the Notch regulating family associated with spondylocostal dysplasia in the human. Studies show, however, that the large majority of human cases with multiple vertebral and rib abnormalities do not demonstrate mutations in these three genes indicating that other mutations almost certainly remain to be discovered for these deformities.

4.4 Genetic Influences on Axial Development: Mutations Identified in Mouse Models with Vertebral Deformation

4.4.1 Hox Genes

Hox genes play a major and very early role in regulating regional development and have been implicated particularly in axial morphogenesis [97–100]. The four mammalian clusters (*Hox A*, *B*, *C*, *and D*) are made up of 39 *Hox* gene family members subdivided into 13 paralogous groups of genes with each group having two to four members. Since there is extensive overlap of expression between the four major adjacent paralogous *Hox* groups, single gene mutations infrequently lead to detectable structural irregularities especially regarding vertebrate morphology, presumably because the remaining and overlapping *Hox* genes involved in any particular segment continue to function normally. Along the vertebrate skeleton from cervical to sacral and tail regions, however, there is an orderly expression of numerical groups of hox genes from *Hox5* to *Hox11*. *Hox* genes elicit their effects as transcription factors. It is also recognized that axial fate is committed very early at the presegmented mesoderm (PSM) stage of development. The identity of a vertebral segment is specified by the combination of functionally active *Hox* genes, and the activity of at least two paralog groups is needed to establish patterning of the axial skeleton. Progressive (numerical) combinations allow for the gradual changing of morphologies as spinal vertebrae are assessed from cranial to caudal regions. *Hox* mutants appear to have greater effects at transition points between vertebral regions rather than within regions. Effects, therefore, are magnified at cervicothoracic, thoracolumbar, and lumbosacral transition levels. Each vertebral element appears to be patterned based on the combination (or overlap) of *Hox* proteins active in each AP region. *Hox5* genes have a sphere of influence in the mid- and lower cervical and upper thoracic region, *Hox6* in the lower cervical and upper and mid-thoracic regions (with the anterior boundaries close to the cervicothoracic junction), *Hox9* in the mid- and lower thoracic and lumbar regions, *Hox 10* in the lumbar and sacral regions (with the anterior limit of gene expression at the boundary between thoracic and lumbar vertebrae), and *Hox11* in the sacral and upper tail regions (with the anterior boundary at the lumbosacral junction). Targeted mutations have been generated in every *Hox* gene in *Hox3* through *Hox11* groups, but the multiple clusters of *Hox* genes that underlie function have led to the need for combining multiple mutations to assess function of vertebrate

Hox genes such that double and triple mutants created experimentally are often needed to elicit regional transformations. Major combinations of *Hox* mutations, however, can have such severe implications that individual embryos may not survive with development abruptly terminated before birth.

4.4.2 Mouse Vertebral Malformations

Several genes specifically involved in early axial segmentation have been identified, and mutations have been detected in one of these described here, pudgy, that are associated with axial malformation leading to vertebral, intervertebral disc, and adjacent rib deformities [109, 117]. Theiler reviewed several hereditary vertebral malformations in mice [16]. Those he defined as disorders of segmentation had the severe and extensive variants of the vertebral and rib abnormalities most similar to the human spondylocostal disorders. These included the pudgy mouse (pu/pu), rib fusions (Rf/Rf), rib-vertebrae (rv/rv), malformed vertebrae (Mv/Mv), Rachiterata (rh/rh), and amputated (am/am), while in one other, crooked tail (Cd/Cd) and lumbosacral abnormalities were seen. Whole mount studies have been performed in the mouse mutant rib-vertebrae (rv/rv) by Nacke et al. [118] showing the vertebral changes, similar to those seen in the pudgy mouse, primarily in the thoracic and lumbar vertebrae along with the characteristic adjacent rib abnormalities. There was defective somite patterning with elongation of the presomitic mesoderm and disruption of the anteroposterior polarization of the somites. They defined abnormal expression of both *Pax1* and *Mox1* along with associated alterations of Notch pathway components Delta-like 1 (*Dll1*) and lunatic fringe (*lfng*). Theiler described many other heritable axial abnormalities in mice but these primarily involved the tail vertebrae and are not directly relevant to the costovertebral disorders we are describing. He also referred to vertebral abnormalities caused by abnormal environmental exposures during the critical days 8–11 of somite formation. Experimentally produced abnormalities were associated with environmental changes such as radiation, anoxia, hypothermia, hyperthermia, hypoglycemia, and several drugs [16]. Sparrow et al. demonstrated that mice with haploinsufficiency of *Hes7* had mild vertebral defects, while wild-type littermates were normal. *Mesp2* +/– embryos had only a few mild vertebral defects. When these strains were exposed to hypoxia in utero, the severity and penetrance of vertebral defects in *Hes7*+/– embryos were increased and a congenital scoliosis-like phenotype was induced in some *Mesp2*+/– embryos [119]. The hypoxia disrupts FGF signaling.

4.4.3 Mutations Affecting the Segmentation Clock and Wavefront: The Molecular Oscillator

Mutations affecting the segmentation clock identified so far in both mouse and human are *Dll3* [2, 107, 109], *Mesp2* [113, 114], *lunatic fringe (lfng)* [116], *Pax1* [120], and *ptk7* [121] as well as *Hes7 (hairy-and-enhancer of split 7)*, *Dusp6*, *Gdf6*, *Gdf3*, *Meox1*, and *Tbx6* [117].

The molecular clock helping to pattern segmentation is activated in the presomitic mesoderm (PSM), the initially non-segmented bilateral mesodermal cell collection running longitudinally in the developing embryo on either side of the neural tube and notochord. It thus affects the developmental process of somitogenesis at its beginning. Following formation of the presomitic mesoderm in which the molecular gradients are established and expressed in cyclic fashion, the bilateral somites form, beginning caudally setting up the pattern of rostro-caudal polarity, such that the first somites come to be positioned anteriorly. The somite borders are formed, differentiation of the various compartments of the somite occurs, and then resegmentation occurs as the vertebral components are organized into their final developmental patterns. Segmentation development in the presomitic mesoderm (PSM) has been closely identified with interplay between Notch, *Wnt3a*, and Fgf8 signaling. The molecular clock operates in the PSM, setting up a series of repetitive molecular gradients. This clock controls the activity of the Notch and Wnt signaling pathways and fibroblast growth factors and disruption of the clock impacts negatively on somite formation and shortly thereafter the segmentation of the sclerotomes. It is the disruption of somitogenesis, due to hereditary disorders to mutations in genes controlling the molecular clock, that induces vertebral and rib malformation syndromes such as the pudgy mouse and its human counterpart spondylocostal dysplasia. There are three gene components of the Notch signaling pathway in which mutations have been identified, in both mouse and human development: Delta-like 3 (*Dll3*), mesoderm posterior 2 (*Mesp2*), and Lunatic fringe (*lfng*). It is expected that mutations in more genes in the molecular oscillatory sequence will be identified and found to underlie generation of vertebral-rib malformation syndromes.

4.4.4 Bovine Vertebral Malformations

A congenital lethal syndrome with marked vertebral and rib malformations, brachyspina syndrome, has been described in Holstein cattle [122–124]. Originally detected in Denmark, cases have since been described in Holland, Italy, and Canada. In this short spine disorder, there are multiple vertebral malformations in 98 %, rib malformations in 94 %, and stiffened tarsal and metatarsal-phalangeal joints in 87 %. Perinatal death is common. The multiple vertebral abnormalities include triangular, wedge-shaped, and fused vertebrae with absent intervertebral discs especially in the cervical and thoracic regions. Ribs are affected being fused, malformed, and bifid proximally adjacent to the vertebrae.

4.5 Congenital Scoliosis (Human): Its Similarity with Pudgy Mouse Vertebral Abnormalities

4.5.1 Incidence and Associated Organ Abnormalities

Congenital scoliosis, specifically defined as a condition due to developmental abnormalities of the vertebral bodies and intervertebral discs, can be estimated to account for approximately 5–10 % of cases of childhood scoliosis in those children in whom a level of deformity warrants formal assessment and management. The overall incidence is not accurately established owing to the fact that many cases are asymptomatic with minimal one or two level abnormalities without angular deformity and detectable only on spine radiographs. A range, however, from 3 to 11 % has been indicated from reviews as far back as the 1930s [125].

The vertebral defects comprising congenital scoliosis can occur as *isolated vertebral and intervertebral disc anomalies* with the patient otherwise well, but there is a high incidence of associated congenital abnormalities in three systems: *neural axis/spinal cord abnormalities* (~40 %), *renal abnormalities* (~25 %), and *cardiac abnormalities* (~10–20 %) [126–131]. The intraspinal abnormalities include diastematomyelia, tethered cord, syringomyelia, Chiari malformation, intraspinal masses (lipoma), and spinal cord malformation with meningocoele, meninogomyelocoele, or rachischisis [132–138]. Renal disorders can include unilateral kidney, renal duplication/ureteral duplication, or ureteral obstruction [128, 139]. Congenital heart defects include mitral valve prolapse, atrial and ventricular septal defects, tetralogy of Fallot, or transposition of the great vessels [127, 138, 140]. Sprengel's deformity (high-riding scapula) is occasionally associated [126]. There is also a syndromal association (other than spondylocostal dysplasia) with the Goldenhar, VACTERL, and Alagille disorders most commonly involved.

4.5.2 Gross Anatomic, Radiologic, and Histopathologic Assessments in Human Congenital Scoliosis

Virtually all reports of the vertebral abnormalities in human congenital scoliosis have been based on gross skeletal and radiologic criteria.

Gross Skeletal Deformities Rokitansky described a clear case of spondylocostal dysplasia in 1839 illustrated with two detailed skeletal illustrations of the spinal and thoracic deformities (scoliosis with wedged and fused vertebrae and rib deformities with asymmetric positioning, decreased numbers, widening and fusions) [3, 141]. Cases of congenital scoliosis using detailed drawings/radiographs/photographs of affected skeletons were described by Putti [4], Werenskiold [142], Feller and Sternberg [143], and Schulz [144] (Fig. 4.3).

Radiologic Examples With the development of radiology, many descriptions followed using either the radiographs or drawings of the radiographs for improved detail. Deformities of the vertebral bodies (congenital scoliosis) were demonstrated by each of Putti [4–6], Fitzwilliams [145], Hodgson [146], Mouchet and Roederer [147], and Sever [148]. In particular, reports by Schulz [144], Hodgson [146], Feller and Sternberg [149], and Putti [5] showed multiple congenital deformities of both vertebrae and ribs accompanied by radiographs (and detailed drawings of the radiographs) to highlight the deformities (Fig. 4.3a1–d2).

Histopathology Few assessments of human congenital scoliosis histopathology have been reported other than small isolated segments in the presentations by Feller and Sternberg in patients with spinal dysraphism (meningomyelocoele and rachischisis) [149, 150], Töndury [78, 79], and Horn et al. assessing the tissue removed at times of surgical corrections [151]. We have presented a detailed histopathologic study of the spine from a child with congenital scoliosis [7] (also included in this paper). Our study involved a radiographic/histopathologic correlation of several (15–16) thoracolumbar levels. Serial sections revealed not only differences between various levels but also differences from anterior to posterior parts of each abnormal vertebral and intervertebral disc segment.

The abnormal radiologic and histologic findings in the pudgy mouse are remarkably similar to those in the severely involved human congenital scoliosis case we have presented. The histologic abnormalities demonstrated are imperfectly sized, shaped, and positioned vertebrae and irregular intervertebral discs. The cartilage and bone, formed by the endochondral sequence or the intramembranous bone formation cascade, when examined as tissues are histologically and histochemically unremarkable. The intervertebral discs are irregular between abnormal vertebrae; they are either thicker or narrower than usual, frequently asymmetric, malpositioned into oblique and even longitudinal orientations, and, where adjacent vertebrae are fused, absent. In some instances, the nucleus pulposus is entirely absent with the interposition only of undifferentiated cartilage between adjacent forming vertebrae. In other areas, the nucleus pulposus is thin or present in two or three separate regions. Invariably, in these situations the annulus fibrosus has failed to form or has formed abnormally. The intervertebral disc abnormalities in association with abnormal bodies are of great interest and must be accounted for in explaining the pathogenesis of the malformations. In viewing radiographs of congenital scoliotic spines, it is important to recognize that the radiolucent areas between abnormal bodies may well represent the cartilage and/or abnormal disc tissue which is more rigid than the normal disc tissue and which may partially or completely ossify with time and growth.

In each of the abnormal vertebral bodies, there is a normal-appearing endochondral sequence except where it has reached the outer margin or periphery of the

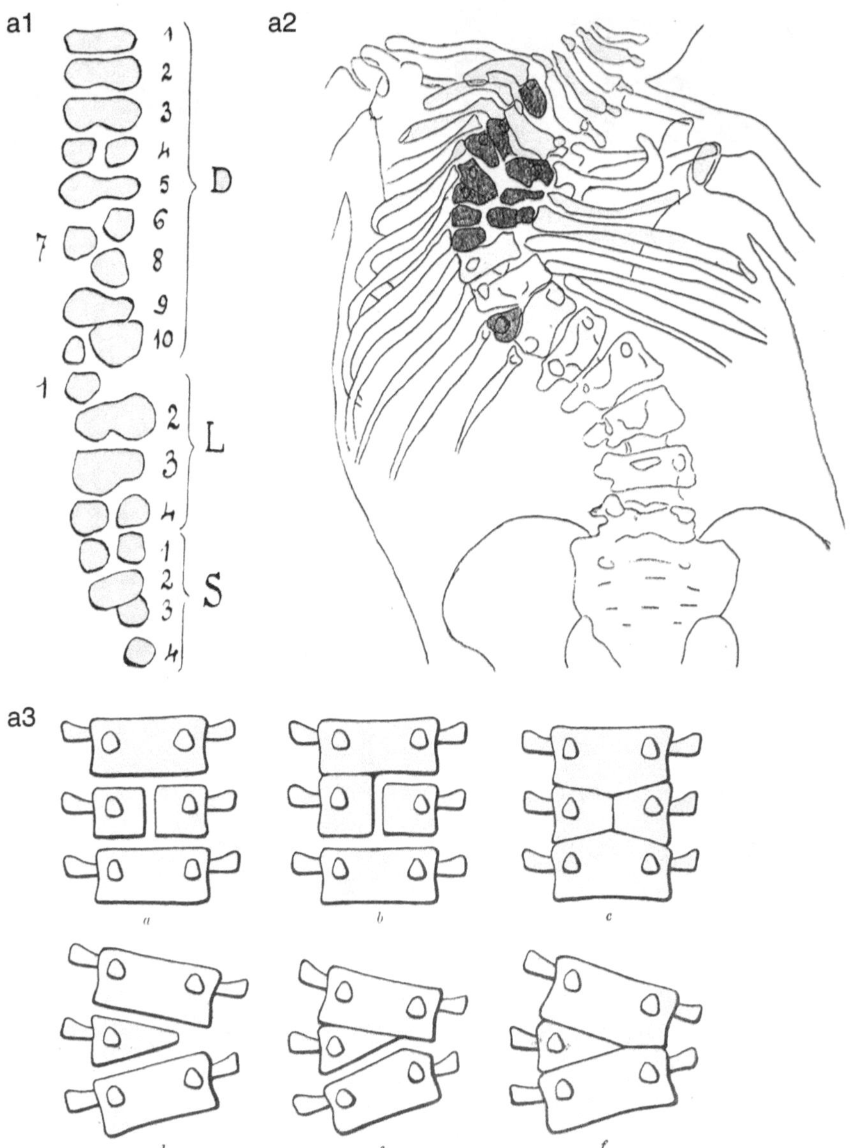

Fig. 4.3 Examples of congenital scoliosis with rib abnormalities (costo-vertebral dysplasias) were encountered and illustrated based on radiographic studies of scoliosis in infancy and childhood in the early twentieth century (**a**1–**a**3). Drawings of deformed spines as revealed on radiographs from the work of Putti show his demonstration of the histopathology of the abnormal vertebrae and adjacent ribs in severe cases. (**b**) Drawing of a radiograph in a case described by Hodgson in 1916 shows multiple vertebral and rib abnormalities. (**c**1, **c**2) Feller and Sternberg described congenital scoliosis with severe rib malformations using detailed drawings of the radiographs (**d**1, **d**2). Schulz described congenital scoliosis and rib abnormalities

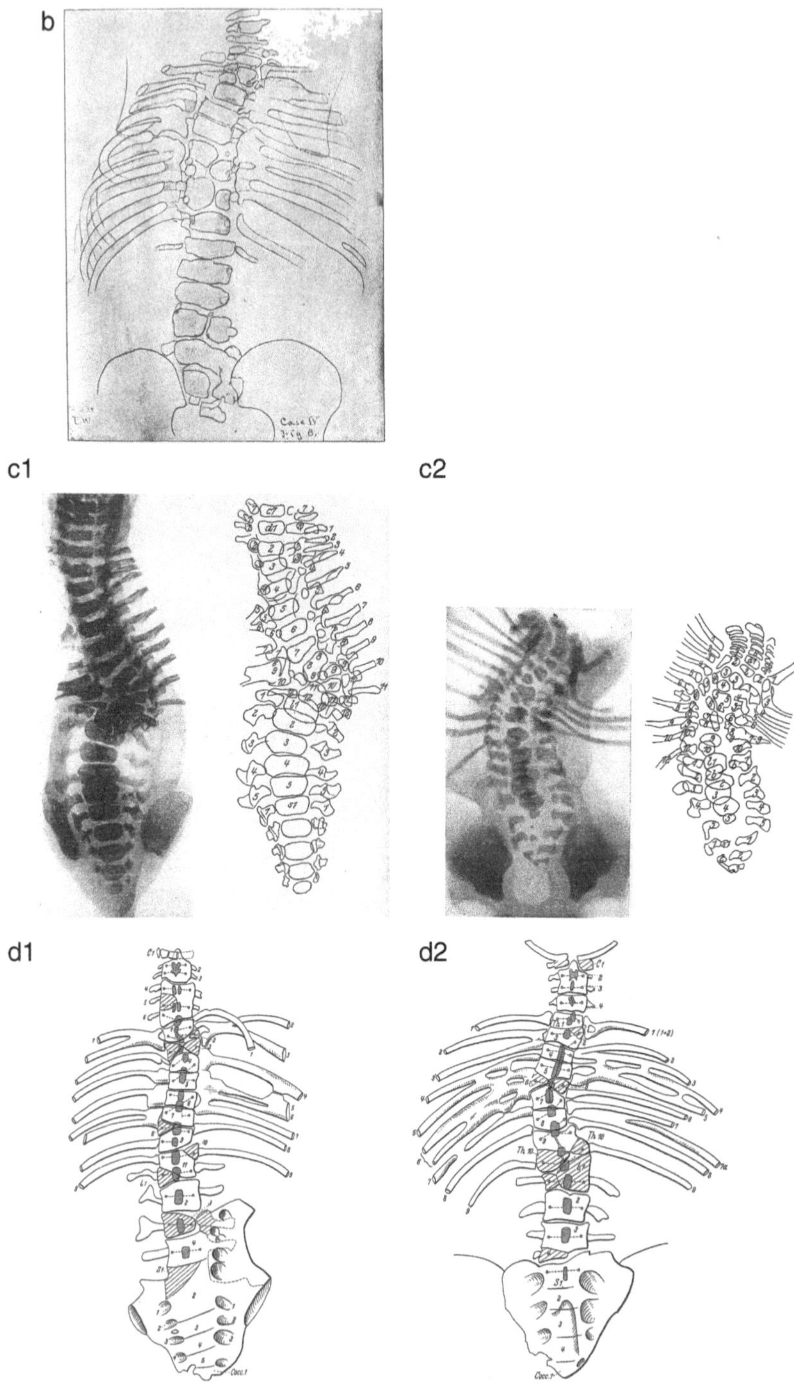

Fig. 4.3 (continued)

cartilage anlage, whereupon normal-appearing intramembranous bone formation occurs. The developing bodies are frequently eccentrically placed in what appears to be an otherwise normal mass of the cartilage tissue. In certain sections, histologically normal intramembranous bone is seen to lie apparently within the body. The serial sections indicate, however, the intramembranous bone is present only in some sections, and one can infer that the outer margin of the cartilage anlage has been reached. Seemingly inappropriate changes into the intramembranous bone thus occur where the cartilage anlage is deficient or where it is present considerably away from the central axis in an area of marked malalignment.

4.5.2.1 Terminology

A brief review of terminology is important as the gene and histomorphologic findings in the various species relate closely to the human clinical disorder, congenital scoliosis. The term *congenital scoliosis* is used in both the mouse model and human clinical realms and refers specifically to structural abnormalities of the vertebrae, intervertebral discs, and ribs present from birth (congenital) and developing from abnormal patterning established at the early embryonic phase and enhanced during fetal growth. This term (congenital scoliosis) has been used clinically at least since the first decade of the twentieth century. Kirmisson described the abnormalities of congenital scoliosis as being associated with bony malformations of the ribs [152]. While most of these deformities in an affected child, or in the mouse, lead to abnormal axial (spinal) curvature in the coronal plane (scoliosis) generally present at birth and worsening with growth, the term (referring to the structurally abnormal vertebrae) is still used in the absence of scoliosis where the various segmental deformities, considered at each spinal level, can on occasion balance each other leaving the spine from cervical region to lumbosacral area straight (normal position). The entity of congenital scoliosis in the human is part of a continuum of early childhood (first decade) spinal deformity (scoliosis) included under the broader term of *early onset scoliosis*. This group encompasses five broad entities: (i) congenital scoliosis (structural), (ii) idiopathic infantile scoliosis (structurally normal spine with a tendency to spontaneous correction or relatively rapid correction with casting, bracing, or physical therapy alone), (iii) neuromuscular scoliosis (due to truncal muscle weakness in such entities as spinal muscular atrophy, myopathies, and congenital muscular dystrophy), (iv) syndromal disorders (often due to connective tissue mutations) such as osteogenesis imperfecta and Marfan syndrome, and (v) spinal tumors. In this work we are discussing only the congenital structural variants. In the orthopedic literature, the term *thoracic insufficiency syndrome* encompasses many of the severe cases of structural congenital scoliosis; it refers to the respiratory compromise associated with thoracic vertebral and rib deformation syndromes that lead to scoliosis and thoracic cage underdevelopment and rigidity.

4.5.3 Description and Categorization of Vertebral and Rib Abnormalities in Congenital Scoliosis

Congenital scoliosis was recognized in the mid-nineteenth century based on the clinical appearance of childhood spinal deformity, often quite severe, and occasional postmortem studies showing both primary vertebral and rib abnormalities. The most severe variants of congenital scoliosis (increasingly referred to as spondylocostal dysplasia or dysostosis) involve structural abnormalities of multiple vertebrae and usually associated abnormalities of the adjacent ribs such as fusions of groups of two to four ribs adjacent to the vertebral column, decreased numbers of ribs, and asymmetric spacing. Rokitansky described and illustrated a case in 1839 [3], as recently pointed out by Nader and Sedivy [141].

With the discovery of X-rays in 1895, radiologic identification of this specific spinal deformity soon outlined its considerable variability ranging from abnormality of a single vertebra to involvement of all vertebrae and many adjacent ribs. Putti wrote a detailed three-part monograph in 1909–1910 outlining the radiographic and morphologic findings [4–6]. He illustrated the radiologic findings and specifically understood the histopathology showing the abnormal shapes of the vertebrae, fusion of adjacent vertebrae, frequent abnormality and even absence of the intervertebral discs, and associated rib fusions (Fig. 4.3a1–a3). He defined five patterns of anomaly from mild to moderate to severe. These were (a) numerical variations of vertebrae within the entire vertebral column, (b) morphologic shape variation of individual vertebrae, (c) numerical plus morphologic variations, (d) faulty differentiation of several adjacent vertebrae, and (e) pathologic malformations [vertebral changes associated with spinal cord abnormalities of myelodysplasia]. Drawings from his work not only show specific deformed vertebrae, for example, wedged and hemivertebrae, but also the variable patterns of the immediately adjacent intervertebral discs ranging from the presence on either side of the abnormal vertebra, to a fusion with the adjacent vertebrae (no disc) on one side and a persisting disc on the other, to complete fusion of the wedged vertebra with vertebrae above and below with no disc interposition (Fig. 4.3a3). Hodgson reported four cases in the American literature in 1916, basing his interpretation on Putti's criteria [146]. One of his illustrations (of a radiograph) shows the rib deformations particularly well (Fig. 4.3b). Lehmann-Facius [153] articulated the view of hemivertebra formation as an error in resegmentation as early as 1925, and Werenskiold [142] discussed radiographic examples of hemivertebrae implicating errors in resegmentation as causing the deformities adding many line drawings to clarify the conception.

4.5.4 Correlation of Clinical, Radiographic, and Histopathologic Findings

Following the extensive investigations by Putti that clearly established the entity, a few studies assessed both the radiologic findings of vertebrae and ribs with a few tissue samples showing the underlying histopathology. Important are those by Feller and Sternberg [143, 149, 150] dealing with vertebral deformities in patients with myelodysplasia and Klippel-Feil disorder (Fig. 4.3c1, c2), Ehrenhaft [74] (showing mostly normal embryo histology), and Töndury [78, 79]. Illustrations from their work highlight the link between the clinical, radiographic, and histopathologic changes. Our detailed study [7], presented in this article as well, correlated all three aspects (Figs. 3.15 and 3.16).

4.5.5 Clinical Recognition of Congenital Scoliosis

Kirmisson in 1910 [152], Klippel and Feil in 1912 [154], Sever in 1922 [148], Mouchet and Roederer in 1922 [147], Kuhns in 1934 [125], and Kuhns and Hormell in 1952 [155] placed the assessment and management of congenital scoliosis into an early clinical framework. The study of the entity was widened when Jarcho and Levin described two cases with extensive vertebral and rib involvement in an article titled "Hereditary malformation of the vertebral bodies" in 1938 [156]. They recognized (i) that malformations of the thoracic and lumbar vertebral bodies were not rare with variations in the number of vertebrae, fusion of two or more adjacent bodies, clefts in the midline, or absence of one-half of a body leaving a hemivertebra and (ii) that cervical vertebral anomalies described by Klippel and Feil in 1912 [154] (leading to a short, rigid neck with low hairline) were a major part of the spectrum of congenital scoliosis. The two cases presented by Jarcho and Levin, which died at 19 days and 6 weeks of age, described more extensive thoracic vertebral and rib malformations leading to serious respiratory symptoms due to the shortening of the spine, thoracic lordosis, and convexity of the anterior chest wall so that "the size of the thoracic cavity is reduced and pulmonary expansion is restrained." The extensive deformities in their cases involved the entire vertebral column but were concentrated in the thoracic and rib regions with some cervical involvement in one case and in the thoracic and rib regions with some lumbar involvement in the other. They presented their cases as an addition to the Klippel-Feil syndrome where the lesions varied "from a cervical spina bifida and simple reduction in the number of vertebrae to an irregular conglomeration or more or less fused vertebral bodies" sometimes extending to the upper thoracic region. In their summary they broadened the entity of vertebral malformation to include (1) malformation of the cervical spine (Klippel-Feil) with a tendency to nonserious symptoms and (2) malformation of the thoracic spine associated with rib malformation where serious respiratory symptoms could compromise life. As subsequent study of congenital axial deformation expanded, cases were variously described as the *Jarcho-Levin syndrome*, *spondylocostal dysplasia or dysostosis*, *spondylothoracic dysplasia*, or *costo-vertebral dysplasia* [157]. [Except when discussing specific articles, we use the term spondylocostal dysplasia (SCD) to refer to the general entity and subhead with gene mutations where identified.]

4.5.6 Classification of Vertebral Disorders

Efforts to classify the human vertebral disorders in the congenital scoliosis spectrum have allowed for a reasonable degree of clinical acceptance. This is primarily because of the fact that, when large numbers of affected patients are considered, most disorders are limited to a few vertebrae involving from one to three segments only which can be encompassed by the categorizations. In classifying congenital scoliosis at a clinical level, abnormalities have been defined as *errors of formation* or *errors of segmentation* based on the understanding that somites form to the right and left sides of the midline in the embryo and with implicit acceptance of the resegmentation mechanism of vertebral body formation. *Errors of formation* are *unilateral complete (=hemivertebra)* or *unilateral partial (=wedged vertebra)*; *errors of segmentation* are *unilateral (=unilateral bar*, involving two or more vertebrae) or *bilateral (=block vertebrae*, involving two or more vertebrae). These are the basic abnormalities with variations occurring, for example, where two or more of the basic patterns occur in the same individual. Winter et al. [158] and McMaster and Ohtsuka [159] outlined these classifications based on their experience with large numbers of cases. The clinical value of these categorizations is that they relate to a fairly large percentage of the abnormalities seen in practice where relatively limited and localized hemivertebrae, wedged vertebrae, unilateral bars, and block vertebrae of two or three levels predominate. On the negative side, as pointed out by Tsou et al. [160] and stressed in this presentation, the relatively simplistic approach does not encompass, let alone begin to explain, the very complex lesions of spondylocostal dysplasia (SCD) in the human, or the pudgy mouse described in this work, where virtually all vertebrae in the spine are affected along with the adjacent ribs. Tsou et al. proposed a descriptive approach that really cannot be called a classification. Attempts have been made recently by a large multispecialty group to address the entire spectrum of disorders encompassing the complex SCD patients with a radiologic classification designed to be further defined by gene abnormalities as they are recognized [161, 162].

4.5.6.1 Congenital Kyphosis

Similar classifications for congenital kyphosis have been developed (Winter et al. [163], McMaster and Singh [164], and Tsou [165]). These encompass errors of formation (with anterior failure of vertebral body formation) leading to a posterior hemivertebra [type I kyphosis, I–A one body, I–B two bodies], while errors of segmentation (anterior failure of segmentation) [type II kyphosis] lead to an anterior bar. Both malformations lead to progressive kyphosis with time as persisting posterior (dorsal) growth outpaces limited or absent anterior (ventral) vertebral body growth. Type III kyphosis is due to a combination of both abnormalities.

4.5.7 Types (Patterns) of Vertebral Deformities

In a report assessing patients with a structural scoliosis from 1908 to 1934, 11 % of the cases (77/681) had a congenital scoliosis (due to abnormalities in the vertebrae and ribs) [125]. The characteristic changes involved fusion of vertebrae, bifid vertebrae, hemivertebrae, and wedge-shaped vertebrae and rib abnormalities with unilateral or bilateral absence, supernumerary ribs, and fused ribs. It was noted "while certain kinds of deformities of the vertebrae and ribs appear again and again, the pattern is never duplicated, and there are no two spines with like deformities in this entire group" [125]. From the same institution 18 years later, the number of cases of congenital scoliosis followed for 3 years or more was 165 [155]. Although rib deformities were found in all cases of extensive congenital scoliosis, the authors mentioned that those deformities "never disturbed thoracic mechanics seriously."

Hemivertebra (usually solitary) is the commonest congenital scoliosis abnormality in large series of patients. In radiographs of 341 patients with significant spinal abnormalities, approximately 40 % had congenital abnormalities; excluding spina bifida occulta which was the most common abnormality (but is rarely of clinical importance), the next commonest were hemivertebrae followed by wedging and then by adjacent vertebral fusions [166]. Hemivertebra was the most numerous variant accounting for 63 of 144 cases (44 %) in one series [160] and 88/251 (35 %) in another large series [159]. In a series of 104 patients with hemivertebrae, 70 (67 %) had a single vertebra, 28 (27 %) two vertebrae, and 6 (6 %) more than two [167]. Touzet et al. reported another large series of 70 cases of hemivertebra [168]. The second most common variant is the unilateral failure of segmentation (unilateral bar); one series documented 171 patients with this deformity with a mean extent of three vertebrae (range, 2–8) [169]. Nasca et al. reported a high incidence of progression of congenital scoliosis (worsening deformity with growth) in particular with hemivertebrae and hemivertebrae with unilateral bars [170]. Based on analysis of multiple studies assessing the entire spectrum of congenital scoliosis, therefore, approximately 60 % of cases involve hemivertebrae (usually single) or unilateral bars (failure of segmentation) with slightly less than 5 % having the multiple vertebrae and severe complex rib anomalies discussed in this study and increasingly referred to as spondylocostal dysplasia (SCD).

Vertebral abnormalities in human congenital scoliosis can be associated with rib/chest wall abnormalities. Tsirikos and McMaster provided the most extensive overview of the frequency of these associations in a large cohort of 620 patients with congenital spine deformities: 497 (80 %) had congenital scoliosis, 88 (14 %) congenital kyphoscoliosis, and 35 (6 %) congenital kyphosis [171]. In the entire series, 119 of 620 patients (19.2 %) had rib abnormalities; most were associated with congenital scoliosis (111 of 497 patients, 22.3 %) with relatively few associated with kyphoscoliosis or kyphosis (8 of 123, 6.5 %). The rib anomalies were simple (one or two ribs involved) in 95 patients and complex (multiple rib fusions, absences) in 24 patients; the 24 patients with complex rib malformations (multiple rib fusions / absences) thus comprised only 4.8 % of congenital scoliosis cases or 4.1 % of the combined scoliosis/kyphoscoliosis cases. Rib deformities were most commonly associated with congenital (thoracic or thoracolumbar) scoliosis in 102/111, 92 % patients and were on the concave side of the congenital scoliosis deformity in 82 (74 %), on the convex side in 22 (20 %), and bilateral in 7 (6 %). The rib abnormalities in congenital scoliosis were heavily concentrated in the unilateral failure of vertebral segmentation (bony bar) patients with 85 of 120 (71 %) affected, while scoliosis due to the commonest deformity, hemivertebra alone, had only 16 of 203 (7.9 %) affected.

Recently, the most complex of the congenital scoliosis disorders, with virtually all vertebrae malformed and multiple rib deformities associated, have become the focus of intense investigation and surgical treatment. These are the inherited spondylothoracic or spondylocostal dysplasias (or dysostoses) [STD/SCD]. These disorders are frequently described as the Jarcho-Levin syndrome [156], but in retrospect were recognized by Rokitansky in 1839 [3, 141] and further defined as specific skeletal dysplasias by Gassner [172] and Roberts et al. [157]. Two large series of cases linked the spondylocostal dysplasias to gene defects in the Notch pathway and also defined the phenotype based on the accumulation of relatively large cohorts of 27 and 13 patients, respectively [173, 174]. The work by Campbell and associates focused on the thoracic deformation in these spondylocostal skeletal dysplasias that progressively limited pulmonary growth and led them to define the *thoracic insufficiency syndrome* [175–177]. The accompanying rib deformations are well illustrated in two of their articles [175, 176]. Scientific interest in this part of the congenital scoliosis spectrum has been enhanced by a combination of (i) surgical intervention on the scoliosis in the first decade of life to straighten the spine without inducing spinal fusion ("growing-rod technique"), (ii) surgical intervention on the rib/thoracic deformities to expand the chest and enhance lung growth (VEPTR—vertical expandable prosthetic titanium rib—technique) [175–178], and (iii) identification of gene abnormalities underlying spondylocostal/spondylothoracic dysplasias such as *Dll3*, *MESP2*, *LNFR* (LUNATIC FRINGE), and others [34, 37, 117, 119, 179–181]. The pudgy mouse and some human spondylocostal dysplasias are both caused by mutations in the delta homologue *Dll3* [2, 107].

4.5.8 Abnormal Human Vertebral Development and Those Parts of the Spectrum Most Closely Aligned with the Pudgy Disorder

The human spinal disorders analogous to the pudgy mice vertebral malformations, where virtually all vertebrae (or at least lengthy segments of vertebrae) along with adjacent ribs are affected, are the skeletal dysplasia disorders referred to as the spondylocostal dysplasias (SCD). This becomes evident from the markedly similar underlying radiographic and histopathologic findings in the pudgy mice and in the human congenital scoliosis case presented. In both the pudgy spine and human congenital scoliosis, the major abnormalities relate (i) to the vertebral bodies which are irregular in size, shape, and position; (ii) to the adjacent intervertebral discs which show varying thicknesses, persistence of cartilage, malpositioned and smaller (though sometimes larger) nuclei pulposi, and imperfect annulus fibrosus formation; and (iii) to the adjacent ribs which show abnormal origins from accumulations of paravertebral cartilage (often encompassing 2–4 vertebrae in length) and are variably fused, branching, absent (deleted), and irregularly spaced. As more cases of the human SCD disorders are assessed by gene analyses, a categorization is now made as follows: SCD I due to mutations of *DLL3*, SCD II due to mutations of *MESP2, and* SCD III due to mutations of *LFNG (lunatic fringe).* It is interesting to note, however, that when large groups of patients with these disorders are screened for these particular gene abnormalities, relatively few have these specific findings. Based on these observations, additional gene defects remain to be detected and defined.

4.5.9 Genes Related to the Klippel-Feil Syndrome

Mutations in *GDF3*, *GDF6*, and *MEOX1* have been shown to underlie the Klippel-Feil syndrome [182]. GDF3 (growth differentiation factor 3) is a member of the TGF-β/BMP (transforming growth factor-beta/bone morphogenetic protein) family and causes the Klippel-Feil 3 deformity. GDF6 (growth differentiation factor 6) is also a member of the TGF-β/BMP family and causes the Klippel-Feil 1 deformity. The protein produced from the *MEOX1* gene (homeobox protein MOX-1) plays a role in somitogenesis and is specifically involved in the formation of the sclerotome [182].

4.6 Pathogenesis of Pudgy and Human Congenital Scoliosis Based on Histopathologic Studies

4.6.1 Vertebral Malformation: Mechanisms Underlying the Development of Structural Abnormalities in Congenital Scoliosis and Descriptive Classifications Based on Them

Possible mechanisms underlying malformation were suggested during the era of radiographic and histopathologic studies several decades before detection of gene abnormalities, and these served as the basis for descriptive categorizations still in use. During the mesenchymal condensation phase, normal segmentation occurs with each somite having a less condensed cranial region and a more condensed caudal region. As development proceeds a hemisegmental shift occurs and it was widely considered that hemivertebrae and/or wedged vertebrae occur because of a failure of both sides to segment in unison. This has been referred to as *faulty unilateral hemimetameric shifting* such that an osseous body is formed only unilaterally [51, 142, 153]. This has an element of descriptive accuracy without providing any insight into causation. It is important to recognize that there is no experimental or investigational evidence for this explanation. Among clinical practice physicians (primarily orthopedic surgeons and radiologists), there is a longstanding descriptive approach dividing the "errors" in axial development into situations that are either *failures of segmentation* or *failures of formation* [158, 159]. Failures of formation refer to *hemivertebra* with complete absence of one-half of the body of a vertebra to either side of the midline, *wedge vertebra* where there is a diminished height of the vertebral body on either side of the midline, or *bifid, double hemivertebra, or a butterfly vertebra* if present on both sides of the midline with central bone continuity absent. Failures of segmentation refer to fused vertebrae where bone continuity between adjacent vertebrae forms instead of intervertebral discs: a *unilateral bar* if present on either side of the midline or a *block vertebra* if across the entire width. This terminology really adds nothing to any understanding of the underlying pathobiology since, in effect, all the disorders can be considered as failures of normal segmentation (or of normal formation considered generally) since the vertebral column is a classic example of formation by repetitive biological segmentation. Accuracy and specificity are further diminished by the use of "mixed types," since differing patterns are seen in many cases and, in the severe variants discussed here, all cases. When essentially all vertebrae and intervertebral discs are abnormal in the severe spondylocostal dysplasia disorders, the formation/segmentation "classification" becomes overwhelmed and ineffective. Even for descriptive purposes, it appears better to describe: (i) the structural findings of the vertebrae, discs, and ribs, (ii) their position and extent, (iii) the degree and plane of angular deformity, and (iv) the gene defect, if defined, without categorization as failures of formation or segmentation.

4.6.2 *Role of Neural Tube/Spinal Cord and Notochord in Development of Axial Deformation: Vertebral Malformation Due to Errors of Induction*

It became well recognized in the 1950s that the developing neural tube/spinal cord and the notochord in embryogenesis influence cartilage differentiation and vertebral morphogenesis of the mesodermal cells surrounding them. Studies were done leading to the conclusion that "it is quite likely that the notochord, as well as the neural tube, possesses the capacity to induce formation of vertebral cartilage" [52]. Earlier work by Lehmann [183] and Kitchen [184] as reported by Watterson et al. [52] had established that elongation of the embryo failed to occur in the absence of the notochord. Experimental studies involving extirpation and/or transplantation of developing regions, primarily the neural tube and notochord, indicated that some form of inducer was needed to allow somites to differentiate into cartilage models. Further studies removing the notochord in various species showed that vertebral cartilage developed but was fused (as a cartilaginous rod) and non-segmented. Holtzer showed in urodele embryos that if a long enough piece of neural tube was removed, the vertebral cartilage failed to form; this indicated a clear role for the neural axis in inducing the vertebral column [53, 54]. Removal of the neural tube led to failure of chondrification and, depending on the sequence, failure of dorsal neural arch formation. When the notochord was removed, however, the response was limited showing its dispensability regarding vertebral cartilage and neural arch formation in urodeles; the notochord actually inhibited cartilage formation in that species. Strudel using the chick embryo worked in this area as well. Removal of early chick neural tube led to failure of formation of the neural arches, while removal of the notochord led to failure of vertebral body formation. Removal of the notochord alone led to a non-segmented cartilage rod ventral to the spinal cord but left the neural arches normal. Strudel summarized clearly: "the nerve tissue (spinal cord) induces differentiation of the dorsal elements of the vertebrae, principally the neural arches. The notochord participates in the differentiation of the ventral elements (......, vertebral centers [bodies], intervertebral discs) and assures the segmentation of the vertebral column"; "the inductive actions of the neural tube and of the notochord are specific" and "in spite of the absence of the notochord, the neural tube or both at the same time .....the limb buds and the ribs develop normally." Watterson et al. added classical studies to chick embryo axial development using both extirpation and transplantation studies of varying combinations of the neural tube and notochord [52]. They showed that complete removal of both the neural tube and notochord prevented the somites from progressing to vertebral cartilage development. Removal of the neural tube only leaving the notochord intact led to formation of centrum-like masses of vertebral cartilage enclosing the notochord. Cartilage formation of the vertebral bodies required the notochord presence and formation of the neural arches required the neural tube presence but the two were independent of each other. Removal of the notochord with the neural tube intact still allowed formation of the vertebral cartilage around the spinal cord, although Strudel had commented that this situation led to non-segmented rods of the cartilage, not true vertebrae. The neural tube was also shown to induce formation of paired neural arches when transplanted into a row of somites. Watterson et al. indicated (similar to Holtzer working with urodeles) that the notochord was to a certain extent dispensable regarding vertebral cartilage formation. However, Watterson et al. showed clearly that the notochord still actively influences vertebral cartilage formation in experiments where it was either transplanted or injured in association with experimental interventions. A transplanted notochord was seen to induce formation of a centrum-like (vertebral body) mass of the cartilage from the host mesoderm. In those situations where the notochord was injured in association with experimental surgeries, it always induced the vertebral body cartilage around itself. This was seen where "each lateral half of the notochord has caused a centrum to form around itself," where the notochord was partially subdivided into three parts, there was excellent correlation between the shape of the centrum and the shape of the modified notochord, and where the notochord was contorted in shape, one could "note conformation between the shape of the centra and shape of the abnormal notochords." In their observations, "production of highly abnormal notochords leads to formation of correspondingly abnormal centra" and "in all cases the size and shape of the centrum conformed to the size and shape of the abnormal notochords to a remarkable degree," these abnormalities being "twin notochords," "tripartite notochords," and "contorted notochords" [52].

It is established that the tissue destined to become vertebral body or disc differentiates in circumferential relationship to the notochord [13]. In separate experiments in chick embryos by Strudel [56, 185, 186] and by Watterson et al. [52] where the notochord was removed, injured or transplanted, development remarkably similar to that seen clinically in congenital scoliosis followed. In both the experimental spines, central cartilage differentiation occurred but normal disc formation and vertebral segmentation did not. Holtzer [53] and Holtzer and Detwiler [54] had previously, using extirpation studies in urodele embryos, shown the negative effects of neural tube removal on vertebral development although notochord extirpation experiments did not invariably lead to vertebral body malformation [53]. The timing of the interventions may be one determining factor since the notochord has its inductive effects very early in development, as early as the presomitic mesoderm stage, and there are clear species differences in responses. Also, the notochord in the chick embryo is proportionally significantly larger in relation to the surrounding cartilage vertebral models than in the mouse or human. The findings of each of these three groups on vertebral development following embryonic neural tube and or notochord extirpation and transplantation, however, are instructive.

For several decades it has been suggested that abnormalities of the notochord might contribute to induction of vertebral body developmental abnormalities. It has been suspected that it is the molecular components of the extracellular matrix of the notochord that carry the inductive principle. These findings emanate from the era when gradients from temporally preceding and spatially adjacent tissues were considered to serve as inducers of subsequent development. The notochord has been frequently considered in relation to vertebral development for three reasons: (i) in vertebrates it forms very early and is positioned centrally in the embryo in relation to both the dorsoventral (DV) and left-right (LR) axes immediately ventral to the neural tube/spinal cord and prior to the development of the sclerotome from which the eventual vertebral bodies differentiate; (ii) in normal development it serves both as an inducing structure to several surrounding organs and as a physical scaffold around which the cartilage and then bone models of the vertebrae form; and (iii) remnants of the notochord contribute to the nucleus pulposus of the intervertebral discs. Continuing investigation of the role of the notochord in embryological development shows that the notochord plays a major role in patterning the neural tube, by signaling the formation of the floor plate, the ventral-most fate of the spinal cord. Notochord signals are also involved in establishing left-right (LR) asymmetry, formation of the dorsal aorta, specification of the cardiac field, and even normal development of the early endoderm and the pancreas [60].

The histological findings in both the pudgy mice and the clinical congenital scoliosis case show deformed, malpositioned, or otherwise imperfect intervertebral discs which are recognized as the remnants of the notochord and are associated with disordered vertebral body development. It is theoretically possible therefore either that: (i) a normal notochord is passively displaced to assume abnormal shapes and positioning by developmental defects originating in the early somites or slightly later sclerotomal cells, (ii) a primarily abnormal notochord (wandering, convoluted, or abnormally shaped) induces an abnormal pattern of mesenchymal or cartilaginous cellular deposition in relation to itself, or (iii) a normal notochord, once passively displaced, enhances the surrounding abnormal cellular development even though it was not directly responsible for originally inducing the abnormal state. Series of extirpation and transplantation experiments in the chick embryo, in urodeles, and in chick-quail chimeras were done several decades ago in efforts to assess the developmental effects of key regions at specific times of development. Töndury commented that abnormal vertebral development invariably follows abnormalities of the notochord [78, 79]. Uhtoff and colleagues on the other hand, based on their interpretation of histologic studies on spontaneously aborted human embryos and fetuses, felt that in spite of findings in experimental studies in lower vertebrates, the notochord did not influence abnormal vertebral body positioning and development [187, 188]. Mutations of genes involved in the oscillatory clock/segmentation mechanism (delta-ll3, Mesp1, lunatic fringe, and others) have been clearly implicated in congenital vertebral and rib malformation. Single gene mutations alone, however, cannot seemingly account for the observations made in this study that all pudgy mice have differing patterns of vertebral, intervertebral disc, and rib deformities. Repetition of the extirpation and transplantation experiments of the notochord and neural tube, in conjunction with assessments by current molecular techniques, should further clarify the exact stages of the abnormal morphogenesis.

4.6.3 Relationship Between Shape and Positioning of the Intervertebral Discs (Nucleus Pulposus) [IVD/NP] and Adjacent Developing Vertebral Bodies in Pudgy Mice

Deviation of the notochord from the midline, a "wandering" or contorted notochord, will encourage the deposition of the mesenchymal tissue in abnormal areas, i.e., malalignment and abnormal shape and size. The notochord early in development could be malpositioned either because it is primarily abnormal itself or it could be normal and secondarily malpositioned simply by relating to the structures developing abnormally due to gene mutations in the segmentation sequence. If the notochord is discontinuous or otherwise abnormal, the intervertebral disc tissue may not be formed, resulting in fused vertebrae, or may be irregularly formed in relation to oblique or vertical positioning leading to hemivertebrae, wedge-shaped vertebrae, or bifid vertebrae. We consider it to be potentially significant that histologic sections of the intervertebral discs (whose nucleus pulposus originated from the notochord and is composed of notochordal remnants) when immediately adjacent to abnormal vertebral bodies are never normal in the pudgy mouse (Figs. 3.8 and 3.11b) or in the case of human congenital scoliosis (Figs. 3.15 and 3.16). We have noted, both in the pudgy mouse histology specimens and in those from the clinical case of congenital scoliosis, that there is invariably a relationship between a structurally abnormal intervertebral disc and an abnormally shaped or positioned vertebral body. Furthermore, the shape and position of the intervertebral disc as viewed in the histologic sections relate specifically to the immediately adjacent malformed vertebral bodies. The more abnormal the intervertebral discs, the more abnormal are the vertebral bodies. Close observations of the abnormal pudgy spine histology indicate that a line extending perpendicular to any section (however short) of the long axis of an abnormal vertebral disc points to the central region of the adjacent bone of the abnormally shaped or positioned developing vertebral body. (These abnormal discs can be in the horizontal, oblique or vertical planes and straight, curved or multisided in shape.) If the disc is positioned in an oblique plane, then the perpendicular line points toward the vertebral body centered to either side of what would have been the normal midline, that is, a wedged, bifid, or hemivertebra body. If the disc is positioned vertically along the long axis of the spine, then the perpendicular line points in both directions (right and left sides) to a bifid vertebra (or two hemivertebrae). These observations are illustrated in Fig. 4.4 where the abnormally positioned and shaped intervertebral discs relate to abnormally positioned vertebral bodies and their ossification sites.

These observations can be stated more formally: *the central region of the cartilage model of a developing vertebral body and its contained endochondral bone lies along a line extending at right angles, or perpendicular, from the central portion of the long axis of the closest section (however short) of the adjacent intervertebral disc (nucleus pulposus) [IVD/NP]*. This general statement pertains

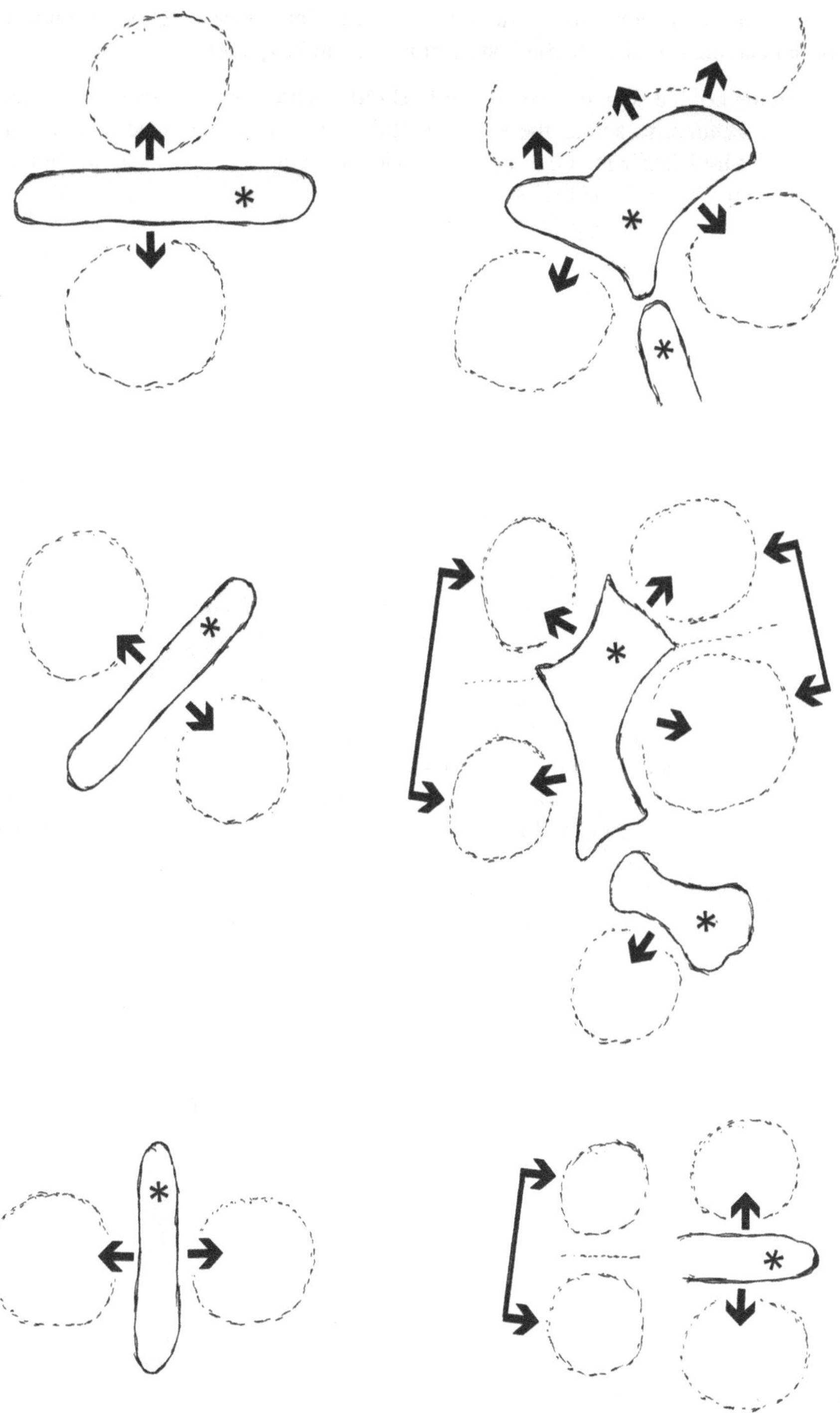

Fig. 4.4 This illustrations outline observations made from multiple histologic sections in non-affected mice (pu/+) and pudgy mice (pu/pu) showing the relationship of normally and

both to normal and abnormal vertebral body formation. Subsequent descriptions of normal and abnormal vertebral body formation can be qualified as follows:

(i) In the normal segment, where the IVD (NP) is horizontally positioned, broad and symmetric across the entire width, it separates two adjacent normally structured cartilage vertebral body models with normal shape and positioning of their central bone regions.
(ii) In a partially imperfectly shaped IVD(NP), where a portion of the IVD (NP) lies in the normal transverse plane, the immediately adjacent vertebral bodies (cartilage model/enclosed bone center) above and below (cranial and caudal directions) are normally structured; where a portion of the disc deviates to an abnormal (e.g., oblique) positioning, the immediately adjacent vertebral body tissue is correspondingly malformed and positioned (lying along a line perpendicular, or at a right angle, to the oblique segment of the disc). The vertebral bone above and below the oblique segment can be continuous with the bone of the normally positioned vertebral body, forming a larger malformed body, or it may be discontinuous forming a separate malformed body (hemivertebra).
(iii) If the abnormal IVD (NP) is vertically aligned, a vertebral body can form on either side (hemivertebra) or on both sides (bifid vertebra) of it, depending on the position of the aberrant disc relative to the midline.
(iv) If the abnormal IVD(NP) is obliquely positioned across the midline, the bone centers of the vertebral bodies above (cranial) and below (caudal) develop perpendicular to the oblique disc and therefore are out of register with one another rather than forming in regular transverse array.
(v) Wherever the IVD (NP) is absent, and therefore not separating the two developing vertebral bodies, the two adjacent cartilage models of the vertebral bodies are in direct apposition to one another; when their bone centers enlarge and reach the edge of the cartilage model with continued growth, bone fusion between the two vertebrae will occur forming a unilateral bone bar (partial disc absence) or block vertebra (complete disc absence).

Based on these findings, concluding observations can be made. *Both the normal and abnormal structure of any vertebral body is directly related in some way to the shape and position of the immediately adjacent intervertebral disc (nucleus pulposus): hemivertebrae and bifid vertebrae form in relation to vertically oriented*

Fig. 4.4 (continued) abnormally positioned and shaped intervertebral discs to immediately adjacent cartilage models and their enclosed bone ossification centers of developing vertebral bodies. These relationships were most clearly observed in the youngest mice assessed from late embryonic and newborn periods. An example of two normal vertebrae and the intervertebral disc between them is seen at the upper right. Each of the other illustrations was adapted from histologic sections from pudgy mice vertebrae. The *arrows* are placed to represent perpendicular axes from the central portion of the adjacent borders of the intervertebral discs to the central portions of the developing bone of the vertebral bodies. The *asterisks* are placed within the intervertebral discs

generally midline discs; wedged vertebrae, and some hemivertebrae, form in relation to obliquely oriented discs; and bony bars (unilateral or block) form in relation to cartilage vertebral body models not separated by discs.

While showing a relationship, this morphologic observation does not indicate the developmental timing of the abnormality or, in particular, which tissue or structural abnormality may have preceded the other. Molecular and structural abnormalities of the notochord could theoretically induce synthesis of the deformed vertebral models primarily. Alternatively, abnormally shaped vertebral bodies positioned and patterned directly by another mechanism (such as the delta-ll3 mutation) could displace the notochord secondarily. Since the young notochord influences the surrounding tissue development, however, even if it has not been primarily displaced and has a normal molecular content, the abnormally positioned notochord still has the potential to potentiate or worsen the position and pattern of vertebral body development.

4.6.4 Stages of Pathogenesis of Spinal Deformity

Based on the observation that no two pudgy mice have the exact same vertebral, disc, and rib abnormalities, it is difficult to attribute the entire pattern of deformation to a single gene defect. For initial or *primary genetic/epigenetic factors* causing deformation, it appears necessary to invoke both the gene mutation affecting the oscillatory segmentation clock and changes induced by epigenetic factors acting along the lines of Waddington's epigenetic landscape. The final deformity is not fully laid out initially. Morphogenesis, either normal or abnormal, is sequential, and following action of the initial primary genetic/epigenetic factors, malformation continues by *secondary obligate mechanisms* (such as defined by positional information) many of which are biologically obligated to occur. For example, the premature deposition of the intramembranous bone is not considered as a primary defect but is rather an "automatic" or obligate response occurring where the endochondral sequence has reached the outer margins of the cartilage anlage or where the anlage is positioned in marked deviation from the midline, even though this is occurring well in advance of the expected time frame. Similarly, in the absence of disc tissue between two adjacent vertebral bodies, once endochondral bone formation in each vertebra reaches the outer rim of the cartilage models, bone tissue forms linking the two bodies into a partial or complete bone fusion. With time, two separate ossification bodies can fuse as their adjacent endochondral sequences, not separated by disc material, grow toward one another. There are even *tertiary biomechanical forces* furthering deformation in the growth period. Once the intramembranous bone has formed in relation to two vertebral bodies, interstitial growth, by which elongation occurs, is no longer possible in that portion of the spine and a tethering effect occurs. If this is unilateral, it leads to progressive curvature of the spine with growth as the persisting cartilage growth regions of the vertebral bodies continue to function. In addition, both wedged and hemivertebrae have increased compressive forces on their concave aspects such that further growth retardation occurs on an added mechanical basis. The normal effects of biophysical forces on musculoskeletal embryologic differentiation and growth were outlined decades ago by Carey in his work on the dynamics of histogenesis ("growth motive force") in the pig [189], and the abnormal forces negatively impacting growth cartilage postnatally, leading to skeletal deformation, were described and formulated in the 1860s as the Hueter-Volkmann law [190, 191]. The manifestations of congenital vertebral deformation in the pudgy mouse or human variants are thus a combination of *primary embryologic abnormalities*, *secondary obligate developmental abnormalities* (such as premature intramembranous bone formation at the periphery of the misaligned or small cartilage anlagen and intervertebral fusion between cartilage regions not separated by normal disc material once their ossification fronts meet each other), and *tertiary biophysical irregularities* (such as increased wedging with time due to increased concave side pressure and decreased convex side pressure which further inhibits or enhances endochondral growth, respectively).

4.6.5 Summary of Rib Malformation

Normal rib development is well illustrated in the non-affected pu/+ specimens. The ribs (as well as the vertebrae) also form by the mechanism of resegmentation. The costo-vertebral relationship is a true synovial joint with a single rib having a double articulation with the vertebral bodies just above and below the intervertebral disc before uniting to form the shaft. A single synovial joint eventually encompasses both heads. The development of the joint is the same as in the large appendicular joints where cells of the articular cartilage surfaces and of the interzone region initially form a continuous cellular mass after which resorption of the interzone hollows out the synovial cavity. Abnormal ribs in severe cases of congenital scoliosis are common. The underlying histopathology of these ribs in either the pudgy mouse or the spondylocostal dysplasias has not been shown previously. Many of the abnormal ribs in the pudgy mice appear to originate from a longitudinal paravertebral series of continuous cartilaginous collections that can be as long as two to four adjacent vertebral bodies. These ossify as the rib ossifies. The cartilage accumulations tend to form some type of articulation with the adjacent vertebral bodies as noted in the younger pudgy mice. We have not noted continuous cartilage bridges from the vertebral bodies to the paravertebral cartilage masses from which the abnormal ribs extend in late embryonic or newborn pudgy mice. There always appears to be either a close to normal synovial joint developing or at least a false articulation, especially in the longer paravertebral longitudinal cartilage accumulations. The whole mount preparations and specimen radiographs often also define the true or aberrant synovial joint structures, but some, on the contralateral side, show what appears to be continuous bone passing from the vertebral body to abnormal ribs. These are seen in the older mice, and it is unclear from our studies whether the costo-vertebral region has gone onto a full bony union or whether the slightly rotated radiologic projection, common with the scoliosis, simply makes it appear that way. As well as forming from the paravertebral cartilage accumulations, there is markedly abnormal patterning with fused ribs leading to widened ribs, branching ribs, asymmetrically positioned ribs, and fewer or deleted ribs. Even the fused ribs, however, elongate via endochondral cartilage mechanisms with shaft formation via the intramembranous bone formation mechanisms.

4.6.6 Summary of Intervertebral Disc Malformation

Intervertebral disc formation is markedly abnormal in pudgy (pu/pu) mice in terms of disc position, shape, size, and even partial or complete absence. The basic pattern of the inner nucleus pulposus and outer annulus fibrosus is maintained, however, although the normal fibrous cross-hatched orientation of the peripheral annulus is also markedly irregular in the abnormal discs. In those places where no intervertebral disc is present between adjacent cartilage vertebral body models, the respective cartilage models are in contact with each other but are distinguishable by cartilage cell flattening at the interface.

4.6.7 Summary of Ganglion Malformation

The study of the pudgy mice ganglia shows clear irregularity of formation. The shape of the involved ganglia is irregular and the ganglionic cell mass appears continuous in the late embryonic and newborn specimens. Closer examination at higher power magnification, however, always reveals thin fibrous septa between or separating adjacent ganglia. Although the appearance of the ganglia is irregular, some of the malformation at least appears to be secondary due to markedly abnormal shapes of the vertebrae and ribs in the immediate area of the ganglia. Of interest also is the observation that at these same time periods in the non-affected (pu/+) mice, the ganglionic tissue is also continuous again showing thin fibrous septa between individual ganglia. This observation is only made when serial sectioning in the normal is continued posteriorly beyond the ribs. Sections within the rib region show the commonly illustrated pattern alternating oval ganglion—rib—oval ganglion—rib, etc. that is the usual way the normal ganglia are portrayed. Continuous ganglionic tissue in the pudgy mouse has been reported previously but there is little awareness that normal ganglionic cells are initially continuous as well. The difference in pudgy from non-affected mice is in the abnormal shaping and positioning of the pudgy ganglia rather than in the early continuity of the cellular tissue.

Chapter 5
Conclusions

The abnormal findings in the pudgy mouse are remarkably similar to those in the severe case of human congenital scoliosis such as seen in the spondylocostal dysplasias. The pudgy mouse is an excellent model to study: (i) how gene mutations translate into three-dimensional structural abnormalities and (ii) which other factors contribute to the final deformities.

5.1 Implications of Pudgy Vertebral Abnormalities for Biologic Research

The variable appearance of the vertebral, disc, and rib abnormalities from mouse to mouse indicates that the *Dll3* gene abnormality is expressed early in embryogenesis in the presomitic mesodermal phase of development but that differing secondary and tertiary effects play a role in outlining the final abnormal patterns. It is increasingly conceivable that cell differentiation decisions mediated by the effects of Waddington's epigenetic landscape lead to differing structural findings in mice with the same gene defects. If the reasonable assumption is made that the same molecular defect is present in each pudgy mouse, the finding that no two mice, even those in the same litter and at the same age, have the exact same patterns of rib or vertebral anomalies leads to the conclusion that solitary gene-molecular defects alone cannot explain all the structural findings. A cascade of subsequent differing multimolecular changes would need to be postulated to explain developmental deformity solely on a molecular basis; therefore high consideration should be given to as yet undefined epigenetic, secondary obligate and tertiary biophysical phenomena which vary subtly from animal to animal once the gene-molecular defect triggers the abnormal sequence. Application of the epigenetic cascade of Waddington is being refined to allow for better understanding of the structural findings in complex anatomic structures. In addition, one can understand how subtle

F. Shapiro, *Disordered Vertebral and Rib Morphology in Pudgy Mice*, Advances in Anatomy, Embryology and Cell Biology 221, DOI 10.1007/978-3-319-43151-2_5

temporal deviations in the oscillation clock can lead to abnormal but differing structural pathways. Abnormal positioning of the intervertebral discs, remnants of the notochord, is demonstrated well in the late embryonic and newborn histology. The concept of a wandering or convoluted notochord as playing a major primary or secondary role in abnormal axial tissue patterning needs to be reconsidered. Even if the notochord is not a primary inciting cause of deformation, its secondary abnormal displacement alone, as early as the presomitic stage, could induce deformation. While morphologic studies alone rarely if ever determine definitive etiology, tracing the histopathology of the earliest embryonic development of all axial structures, as well as those originating from the presomitic mesoderm, would give clearer signals as to pathogenesis. Whole mount preparations, specimen radiographs, histologic studies, and three-dimensional computerized reconstructions of histologic sections of normal and pudgy spines all depict well normal and abnormal structuring in the pudgy mouse. With reconstructions of axial regions, we can show each of the individual components, any combination of the individual components, the individual components in 360° rotation, and cutaways to allow for an "internal" view. Tissue reconstructions are done more easily and with almost the same resolution with magnetic resonance imaging; the reconstructions described in this paper were based on a technique designed to assess reconstructions at the cellular level.

5.2 Implications of Pudgy Vertebral Abnormalities for Clinical Patient Treatment

Congenital scoliosis is very difficult to treat since the deformities are present from the early embryonic period. The vertebral deformities can be markedly asymmetric, often leading to scoliotic deformity evident at birth that tends not to respond well to bracing owing to the rigidity of the spine occasioned by the structurally abnormal intervertebral discs and the consequent lack of flexibility for better spinal positioning. Growth tends to exacerbate the deformity by *secondary obligate* and *tertiary mechanical* imbalance. Associated rib abnormalities, which often involve asymmetric fusion of series of ribs, further alter growth, restrict respiratory function, and limit application of bracing which itself diminishes respiratory function with any meaningful force applied. Management could possibly be improved by better awareness of the structural abnormalities in the early phases when developing cartilage and intervertebral disc tissues are still radiolucent (and thus non-detectable by plain radiographs). Bone tissue can currently be defined by plain radiographs and computerized axial tomography scanning, but better use of magnetic resonance imaging can identify each of the cartilage, nucleus pulposus, and annulus fibrosus positioning which might be amenable to more select surgical intervention.

Acknowledgments We thank Lauren Barone, George Malatantis, Christopher Cahill, and Evelyn Flynn for excellent technical help and "Advanced Medical Graphics" of Boston, MA, for excellent assistance in the presentation of the figures.

References

1. Grünberg H (1961) Genetical studies on the skeleton of the mouse. XXIX Pudgy Genet Res 2:384–393
2. Kusumi K, Sun ES, Kerrebrock AW, Bronson RT, Chi D-C, Bulotsky MS, Spencer JB, Birren BW, Frankel WN, Lander ES (1998) The mouse pudgy mutation disrupts *Delta* homologue *Dll3* and initiation of early somite boundaries. Nat Genet 19:274–278
3. Rokitansky C (1839) Beyträge zur Kenntnis der Rückgrathskrümmungen, und der mit denselben zusammentreffenden Abweichungen des Brustkorbes und Beckens. Schluss Medizinisches Jahrbuch des k k österreichischen Staates. NF XXVIII(28):195–211
4. Putti V (1909) Die Angeborenen Deformitäten der Wirbelsäule. Fortschr Röntgenstr 14:285–313
5. Putti V (1910) Die Angeborenen Deformitäten der Wirbelsäule. Fortschr Röntgenstr 15:65–92
6. Putti V (1910) Die Angeborenen Deformitäten der Wirbelsäule. Fortschr Röntgenstr 15:243–292
7. Shapiro F, Eyre DR (1981) Congenital scoliosis. A histopathologic study. Spine 6:107–118
8. Lundvall H (1904) Über Demonstration embryonaler Knorpelskelette. Anat Anz 25:219–222
9. Lundvall H (1905) Weiteres über Demonstration embryonaler Skelette. Anat Anz 27:520–523
10. Simons EV, Van Horn JR (1971) A new procedure for whole mount alcian blue staining of the cartilaginous skeleton of chicken embryos, adapted to the clearing procedure in potassium hydroxide. Acta Morphol Neerl-Scand 8:281–292
11. Harris KM, Stevens JK (1989) Dendritic spines of CA 1 pyramidal cells in the rat hippocampus: serial electron microscopy with reference to their biophysical characteristics. J Neurosci 9(8):2982–2997
12. Harris KM, Jensen FE, Tsao B (1992) Three-dimensional structure of dendritic spines and synapses in rat hippocampus (CA1) at postnatal day 15 and adult ages: implications for the maturation of synaptic physiology and long term potentiation. J Neurosci 12(7):2685–2705
13. Shapiro F (1992) Vertebral development of the chick embryo during days 3-19 of incubation. J Morphol 213:317–333
14. Dawes B (1930) The development of the vertebral column in mammals as illustrated by its development in Mus Musculus. Philos Trans R Soc Lond Ser B 218:115–170
15. Theiler K (1989) The house mouse. Atlas of embryonic development. Springer, Berlin, New York
16. Theiler K (1988) Vertebral malformations. Adv Anat Embryol Cell Biol 112:1–99

F. Shapiro, *Disordered Vertebral and Rib Morphology in Pudgy Mice*, Advances in Anatomy, Embryology and Cell Biology 221, DOI 10.1007/978-3-319-43151-2

17. Von Baer KE (1828) Über Entwicklungsgeschichte der Thiere. Beobachtung und Reflexion, Königsberg
18. Remak R (1855) Untersuchungen über die Entwicklung der Wirbelthiere. Reimer, Berlin
19. Von Ebner V (1888) Urwirbel und Neugliederung der Wirbelsäule. Sitzungsber Akad Wiss Wien 97:194–206
20. Crick FHC, Lawrence PA (1975) Compartments and polyclones in insect development. Science 189:340–347
21. Wolpert L (1969) Positional information and the spatial pattern of cellular differentiation. J Theoret Biol 25:1–47
22. Moore WJ, Mintz B (1972) Clonal model of vertebral column and skull development derived from genetically mosaic skeletons in allophenic mice. Dev Biol 27:55–70
23. Waddington CH (1942) Canalization of development and the inheritance of acquired characters. Nature 150:563–565
24. Waddington CH (1957) The strategy of the genes. A discussion of some aspects of theoretical biology. George Allen and Unwin, London
25. Pourquié O (2003) The segmentation clock: converting embryonic time into spatial pattern. Science 301:328–330
26. Bhattacharya S, Zhang Q, Andersen ME (2011) A deterministic map of Waddington's epigenetic landscape for cell fate specification. BMC Syst Biol 5:85
27. Wang J, Zhang K, Xu L, Wang E (2011) Quantifying the Waddington landscape and biological paths for development and differentiation. Proc Natl Acad Sci USA 108:8257–8262
28. Goldberg AD, Allis CD, Bernstein E (2007) Epigenetics: a landscape takes shape. Cell 128:635–638
29. Cooke J, Zeeman EC (1976) A clock and wavefront model for control of the number of repeated structures during animal morphogenesis. J Theoret Biol 58:455–476
30. Kmita M, Duboule D (2003) Organizing axes in time and space; 25 years of collinear tinkering. Science 301:331–333
31. Sparrow DB, Chapman G, Turnpenny PD, Dunwoodie SL (2007) Disruption of the somatic molecular clock causes abnormal vertebral segmentation. Birth Defects Res C Embryo Today 81:93–110
32. Sewell W, Kusumi K (2007) Genetic analysis of molecular oscillators in mammalian somitogenesis. Birth Defects Res C Embryo Today 81:111–120
33. Shifley ET, Cole SE (2007) The vertebrate segmentation clock and its role in skeletal defects. Birth Defects Res C Embryo Today 81:121–133
34. Pourquié O (2011) Vertebrate segmentation: from cyclic gene networks to scoliosis. Cell 145:650–663
35. Eckalbar WL, Fisher RE, Rawls A, Kusumi K (2012) Scoliosis and segmentation defects of the vertebrae. WIREs Dev Biol 1:401–423
36. Turnpenny PD, Alman B, Cornier AS, Giampietro PF, Offiah A, Tassy O, Pourquie O, Kusumi K, Dunwoodie S (2007) Abnormal vertebral segmentation and the notch signaling pathway in man. Dev Dynam 236:1456–1474
37. Sparrow DB, Chapman G, Dunwoodie SL (2011) The mouse notches up another success: understanding the causes of human vertebral malformation. Mamm Genome 22:362–376
38. Aulehla A, Wehrle C, Brand-Saberi B, Kember R, Gossler A, Kanzler B, Hermann BG (2003) *Wnt3a* plays a major role in the segmentation clock controlling somitogenesis. Dev Cell 4:395–406
39. Baker RE, Schnell S, Maini PK (2006) A clock and wavefront mechanism for somite formation. Dev Biol 293:116–126
40. Keynes RJ, Stern CD (1988) Mechanisms of vertebrate segmentation. Development 103:413–429
41. Verbout AJ (1985) The development of the vertebral column. Adv Anat Embryol Cell Biol 90:1–122

42. Fleming A, Kishida MG, Kimmel CB, Keynes RJ (2015) Building the backbone: the development and evolution of vertebral patterning. Development 142:1733–1744
43. Christ B, Wilting J (1992) From somites to vertebral column. Ann Anat 174:23–32
44. Christ B, Ordahl CP (1995) Early stages of chick somite development. Anat Embryol 191:381–396
45. Christ B, Huang R, Wilting J (2000) The development of the avian vertebral column. Anat Embryol 202:179–194
46. Stockdale FE, Nikovits W Jr, Christ B (2000) Molecular and cellular biology of avian somite development. Dev Dynam 219:304–321
47. Christ B, Huang R, Scaal M (2007) Amniote somite derivatives. Dev Dynam 236:2382–2396
48. Bagnall KM, Higgins SJ, Sanders EJ (1988) The contribution made by a single somite to the vertebral column: experimental evidence in support of resegmentation using the chick- quail chimaera model. Development 103:69–85
49. Nowicki JL, Takimoto R, Burke AC (2003) The lateral somite frontier: dorso-ventral aspects of anterior-posterior regionalization in avian embryos. Mech Dev 120:227–240
50. Spörle R (2001) Epaxial-adaxial-hypaxial regionalisation of the vertebrate somite: evidence for a somitic organizer and a mirror-image duplication. Dev Genes Evol 211:198–217
51. Hall BK (1977) Chondrogenesis of the somitic mesoderm. Adv Anat Embryol Cell Biol 53:1–49
52. Watterson RL, Fowler I, Fowler BJ (1954) The role of the neural tube and notochord in development of the axial skeleton of the chick. Am J Anat 95:337–399
53. Holtzer H (1952) An experimental analysis of the development of the spinal column. Part II. The dispensability of the notochord. J Exp Zool 121:573–591
54. Holtzer H, Detwiler SR (1953) An experimental analysis of the development of the spinal column. III. Inductionof Skeletogenous cells. J Exp Zool 123:335–369
55. Corallo D, Trapani V, Bonaldo P (2015) The notochord: structure and functions. Cell Mol Life Sci 72:2989–3008
56. Strudel G (1955) L'action morphogène du tube nerveux et de la corde sur la différentiation des vertèbres et des muscles vertébraux chez l'embryon de poulet. Arch d'Anat Microscop Morph Exp 44:209–235
57. Yamada T, Placzek M, Tanaka H, Dodd J, Jessell TM (1991) Control of cell pattern in the developing nervous system: polarizing activity of the floor plate and notochord. Cell 64:635–647
58. Goldstein RS, Kalcheim C (1991) Normal segmentation and size of the primary sympathetic ganglia depend upon the alteration of rostrocaudal properties of the somites. Development 112:327–334
59. Anderson CNG, Ohta K, Quick MM, Fleming A, Keynes R, Tannahill D (2003) Molecular analysis of axon repulsion by the notochord. Development 130:1123–1133
60. Stemple DL (2005) Structure and function of the notochord: an essential organ for chordate development. Development 132:2503–2512
61. Jurand A (1974) Some aspects of the development of the notochord in mouse embryos. J Embryol Exp Morph 32:1–33
62. Pourquie O, Coltey M, Teillet M-A, Ordahl C, LeDouarin NM (1993) Control of dorsoventral patterning of somitic derivatives by notochord and floor plate. Proc Natl Acad Sci USA 90:5242–5246
63. Williams LW (1908) The later development of the notochord in mammals. Am J Anat 8:251–283
64. Duval M (1889) Atlas d'Embryologie. G. Masson, Paris
65. Romanoff AL (1960) The skeletal system and the integument. In: Avian embryo. Structural and functional development. MacMillan Company, New York, pp 907–1007
66. Arey LB (1965) The study of chick embryos, chapter 28. In: Developmental anatomy. A textbook and laboratory manual of embryology, 7th edn. WB Saunders, Philadelphia, pp 551–609

67. Tam PPL, Trainor PA (1994) Specification and segmentation of the paraxial mesoderm. Anat Embryol 189:275–305
68. Williams LW (1910) The somites of the chick. Am J Anat 11:55–100
69. Verbout AJ (1976) A critical review of the 'Neugliedererung' concept in relation to the development of the vertebral column. Acta Biotheor 25:218–258
70. Baur R (1969) Zum Problem der Neugliederung der Wirbelsäule. Acta Anat 72:321–356
71. Dalgleish AE (1985) A study of the development of thoracic vertebrae in the mouse assisted by autoradiography. Acta Anat 122:91–98
72. Piiper J (1928) On the evolution of the vertebral column in birds, illustrated by its development in Larus and Struthio. Philos Trans R Soc Lond B 216:285–351
73. Bardeen CR (1905) The development of the thoracic vertebrae in man. Am J Anat 4:163–174
74. Ehrenhaft JL (1943) Development of the vertebral column as related to certain congenital and pathological changes. Surg Gynecol Obstet 76:282–292
75. Wyburn GM (1944) Observations on the development of the human vertebral column. J Anat 78:94–102
76. Sensenig EC (1949) The early development of the human vertebral column. Contrib Embryol 33:23–41
77. Bick EM, Copel JW (1950) Longitudinal growth of the human vertebra. J Bone Joint Surg Am 32A:803–814
78. Töndury G (1952) Neuere Ergebnisse über die Entwickslungsphysiologie der Wirbelsäule. Arch Orthop Unfall-Chir 45:313–22
79. Töndury G (1953) Le développement de la colonne vertébrale. Rev Chir Orthop 39:553–569
80. O'Rahilly R, Meyer DB (1979) The timing and sequence of events in the development of the human vertebral column during the embryonic period proper. Anat Embryol 157:167–176
81. Keyes DC, Compere EL (1932) The normal and pathological physiology of the nucleus pulposus of the intervertebral disc. J Bone Joint Surg 14:897–938
82. Walmsley R (1953) The development and growth of the intervertebral disc. Edinburgh Med J 60:341–364
83. Roberts S, Evans H, Trivedi J, Menage J (2006) Histology and pathology of the human intervertebral disc. J Bone Joint Surg Am 88(Suppl 2):10–14
84. Gadow H, Abbott EC (1895) On the evolution of the vertebral column of fishes. Philos Trans Ser B 186:163–221
85. Gadow H (1896) On the evolution of the vertebral column of amphibian and amniota. Philos Trans Ser B 187:1–57
86. Williams EE (1959) Gadow's arcualia and the development of tetrapod vertebrae. Quart Rev Biol 34:1–32
87. Hautier L, Charles C, Asher RJ, Gaunt SJ (2014) Ossification sequence and genetic patterning in the mouse axial skeleton. J Exp Zool B Mol Dev Evol 322:631–642
88. Bagnall KM, Harris PF, Jones RPM (1977) A radiographic study of the human fetal spine. 2. The sequence of development of ossification centres in the vertebral column. J Anat 124:791–802
89. O'Rahilly R, Muller F, Meyer DB (1980) The human vertebral column at the end of the embryonic period proper. 1. The column as a whole. J Anat 131:565–575
90. Williams JL (1942) The development of cervical vertebrae in the chick under normal and experimental conditions. Am J Anat 71:153–179
91. Christ B, Jacob HJ, Jacob M (1974) Experimentelle Untersuchungen zur Entwicklung der brustwand beim Hühnerembryo. Experientia 30:1449–1451
92. Kato N, Aoyama H (1998) Dermomyotomal origin of the ribs as revealed by extirpation and transplantation experiments in chick and quail embryos. Development 125:3437–3443
93. Huang R, Zhi Q, Schmidt C, Wiltig J, Brand-Saberi B, Christ B (2000) Sclerotomal origin of the ribs. Development 127:527–532
94. Aoyama H, Mizutani-Koseki Y, Koseki H (2005) Three developmental compartments involved in rib formation. Int J Dev Biol 49:325–333

95. Burke AC, Nowicki JL (2003) A new view of patterning domains in the vertebrate mesoderm. Dev Cell 4:159–165
96. Evans DJR (2003) Contribution of somatic cells to the avian ribs. Dev Biol 256:114–126
97. Wellik DM (2007) *Hox* patterning of the vertebral axial skeleton. Dev Dynam 236:2454–2463
98. Wellik DM (2009) *Hox* genes and vertebrate axial pattern. Curr Top Dev Biol 88:257–278
99. Mallo M, Wellik DM, Deschamps J (2010) *Hox genes* and regional patterning of the vertebrate body plan. Dev Biol 344:7–15
100. McIntyre DC, Rakshit S, Yallowitz AR, Loken L, Jeannotte L, Capecchi MR, Wellik DM (2007) Hox patterning of the vertebrate rib cage. Development 134:2981–2989
101. Bick EM, Copel JW (1951) The ring apophysis of the human vertebra. J Bone Joint Surg 33:783–787
102. Dahia CL, Mahoney EJ, Durrani AA, Wylie C (2009) Post natal growth, differentiation, and aging of the mouse intervertebral disc. Spine 34:447–455
103. Zhang Y, Lenart BA, Lee JK, Chen D, Shi P, Ren J, Muehlman C, Chen D, An HS (2014) Histological features of the endplates of the mammalian spine: from mice to men. Spine 39: E312–E317
104. Hamburger V, Hamilton HL (1951) A series of normal stages in the development of the chicken embryo. J Morph 88:49–92
105. Bancroft M, Bellairs R (1976) The development of the notochord in the chick embryo studied by scanning and transmission electron microscopy. J Embryol Exp Morphol 35:383–401
106. Turnpenny PD, Bulman MP, Frayling TM, Abu-Nasra TK, Garrett C, Hattersley AT (1999) A gene for autosomal recessive spondylocostal dysostosis maps to 19q 13.1-q 13.3. Am J Hum Genet 65:175–182
107. Bulman MP, Kusumi K, Frayling TM, McKeown C, Garrett C, Lander ES, Krumlauf R, Hattersley AT, Ellard S, Turnpenny PD (2000) Mutations in the human delta homologue, DLL3, cause axial skeletal defects in spondylocostal dysostosis. Nat Genet 24:438–441
108. Dunwoodie SL, Clements M, Sparrow DB, Sa X, Conlon RA, Beddington RSP (2002) Axial skeletal defects caused by mutation in the spondylocostal dysplasia/pudgy gene *Dll3* are associated with disruption of the segmentation clock within the presomitic medsoderm. Development 129:1795–1806
109. Kusumi K, Mimoto MS, Covello KL, Beddington RSP, Krumlauf R, Dunwoodie SL (2004) *Dll3* pudgy mutation differentially disrupts dynamic expression of somite genes. Genesis 39:115–21
110. Saga Y, Hata N, Koseki H, Taketo MM (1997) *Mesp2*: a novel mousegene expressed in the presegmented mesoderm and essential for segmentation initiation. Genes Dev 11:1827–1839
111. Takahashi Y, Koizumi K, Takagi A, Kitajima S, Inoue T, Koseki H, Saga Y (2000) *Mesp2* initiates somite segmentation through the Notch signaling pathway. Nat Genet 25:390–396
112. Morimoto M, Takahashi Y, Endo M, Saga Y (2005) The *Mesp2* transcription factor establishes segmental borders by suppressing Notch activity. Nature 435:354–359
113. Whittock NV, Sparrow DB, Wouters MA, Sillence D, Ellard S, Dunwoodie SL, Turnpenny PD (2004) Mutated *MESP2* causes spondylocostal dysostosis in humans. Am J Hum Genet 74:1249–1254
114. Cornier AS, Staehling-Hampton K, Delventhal KM, Saga Y, Caubet J-F, Sasaki N, Ellard S, Young E, Ramirez N, Carlo SE, Torres J, Emans JB, Turnpenny PD, Pourquié O (2008) Mutations in the *MESP2* gene cause spondylothoracic dysostosis/Jarcho-Levin syndrome. Am J Hum Genet 82:1334–1341
115. Evrard YA, Lun Y, Aulehla A, Gan L, Johnson RL (1998) *lunatic fringe* is an essential mediator of somite segmentation and patterning. Nature 394:377–381
116. Sparrow DB, Chapman G, Wouters MA, Whittock NV, Ellard S, Fatkin D, Turnpenny PD, Kusumi K, Sillence D, Dunwoodie SL (2006) Mutation of the *LUNATIC FRINGE* gene in humans causes spondylocostal dysostosis with a severe vertebral phenotype. Am J Hum Genet 78:28–37

117. Dunwoodie SL, Sparrow DB (2015) Genetic and environmental interaction in the malformation of the vertebral column. In: Wise CA, Rios JJ (eds) Molecular genetics of pediatric orthopedic disorders. Springer Science + Business Media, New York, pp 131–51
118. Nacke S, Schäfer R, Hrabé de Angelis M, Mundlos S (2000) Mouse mutant "rib-vertebrae" (rv): a defect in somite polarity. Dev Dynam 219:192–200
119. Sparrow DB, Chapman G, Smith AJ, Mattar MZ, Major JA, O'Reilly VC, Saga Y, Zackai EH, Dormans JP, Alman BA, McGregor L, Kageyama R, Kusumi K, Dunwoodie SL (2012) A mechanism for gene-environment interaction in the etiology of congenital scoliosis. Cell 149:295–306
120. Giampietro PF, Raggio CL, Reynolds CE, Shukla SK, McPherson E, Ghebranious N et al (2005) An analysis of *PAX1* in the development of vertebral malformations. Clin Genet 68:448–453
121. Hayes M, Gao X, Yu LX, Paria N, Henkelman RM, Wise CA, Ciruna B (2014) *ptk7* mutant zebrafish models of congenital and idiopathic scoliosis implicate dysregulated Wnt signaling in disease. Nat Commun 5:4777. doi:10.1038/ncomms5777
122. Agerholm JS, McEvoy F, Arnbjerg J (2006) Brachyspina syndrome in a Holstein calf. J Vet Diagn Invest 18:418–422
123. Agerholm JS (2007) Inherited disorders in Danish cattle. APMIS 115(Suppl 122):1–76
124. Agerholm JS, DeLay J, Hicks B, Fredholm M (2010) First confirmed case of the bovine brachyspina syndrome in Canada. Can Vet J 51:1349–1350
125. Kuhns JG (1934) Congenital scoliosis. N Engl J Med 210:1310–1315
126. Beals RK, Robbins JR, Rolfe B (1993) Anomalies associated with vertebral malformations. Spine 18:1329–1332
127. Basu PS, Elsebaie H, Noordeen MHH (2002) Congenital spinal deformity: a comprehensive assessment at presentation. Spine 27(20):2255–2259
128. Debnath UK, Goel V, Harshavardhana N, Webb JK (2010) Congenital scoliosis-Quo vadis? Indian J Orthop 44:137–147
129. Arlet V, Odent T, Aebi M (2003) Congenital scoliosis. Eur Spine J 12:456–463
130. Hensinger RN (2009) Congenital scoliosis: etiology and associations. Spine 34 (17):1745–1750
131. Kaspiris A, Grivas TB, Weiss H-R, Turnbull D (2011) Surgical and conservative treatment of patients with congenital scoliosis: a search for long-term results. Scoliosis 6:12
132. Prahinski JR, Polly DW Jr, McHale KA, Ellenbogen RG (2000) Occult intraspinal anomalies in congenital scoliosis. J Pediatr Orthop 20:59–63
133. Belmont PJ Jr, Kuklo TR, Taylor KF, Freedman BA, Prahinski JR, Kruse RW (2004) Intraspinal anomalies associated with isolated congenital hemivertebrae: the role of MRI. J Bone Joint Surg Am 86:1704–1710
134. Suh SW, Sarwark JF, Vora A, Huang BK (2001) Evaluating congenital spine deformities for intaspinal anomalies with magnetic resonance imaging. J Pediatr Orthop 21:525–531
135. Pigott H (1980) The natural history of scoliosisi in myelodysplasia. J Bone Joint Surg Br 62:54–58
136. Raycroft JF, Curtis BH (1972) Spinal curvature in myelomeningocele: natural history and etiology. In: Proceedings American academy of orthopaedic surgeons. Symposium on myelomeningocele. St. Louis, CV Mosby, pp 186–201
137. Winter RB, Haven JJ, Moe JH, Lagaard SM (1974) Diastematomyelia and congenital spine deformities. J Bone Joint Surg Am 56:27–39
138. Liu Y, Guo L, Tian Z, Zhu W, Yu B, Zhang S, Qiu G (2011) A retrospective study of congenital scoliosis and associated cardiac and intraspinal abnormalities in a Chinese population. Eur Spine J 20:2111–2114
139. Drvaric DM, Ruderman RJ, Coonrad RW, Grossman H, Webster GD, Schmitt EW (1987) Congenital scoliosis and urinary tract abnormalities: are intravenous pyelograms necessary? J Pediatr Orthop 7:441–443

140. Reckles L, Peterson H, Bianco A, Weidman W (1975) The association of scoliosis and congenital heart defects. J Bone Joint Surg Am 57:449–455
141. Nader A, Sedivy R (2004) Rokitanskys Erstbeschreibung einer spondylokostalen Dysplasie [Rokitansky's first description of a spondylocostal dysplasia, dysostosis]. Wien Med Wochenschr 154:472–474
142. Werenskiold B (1937) Über einen fall von Wirbelmissbildungen, Keilwirbel, Spiralwirbel. Acta Radiol os-18:775–97
143. Feller A, Sternberg H (1932) Zur Kenntnis der Fehlbildungen der Wirbelsäule. IV. Die anatomischen Grundlagen des Kurzhalses (Klippel-Feilschen Syndroms). Virchows Archiv f Path Anat Physiol 285:112–139
144. Schulz O (1939) Beitrage zur Kenntnis der Fehlbildungen der Wirbelsäule. Virchows Archiv f Path Anat Physiol 304:203–222
145. Fitzwilliams DCL (1909) Case of congenital scoliosis [Section, study diseases of children]. Proc R Soc Med 2:25–33
146. Hodgson FG (1916) Congenital deformities of the vertebrae and ribs. A report of four cases. J Bone Joint Surg 14(s2):34–44
147. Mouchet A, Roederer C (1922) Considerations sur pathogenie et l'evolution de la scoliose congénitale. La Presse Med #54, 8 juillet:577–579
148. Sever JW (1922) Congenital anatomic defects of the spine and ribs. Boston Med Surg J 186:799–821
149. Feller A, Sternberg H (1930) Zur Kenntnis der Fehlbildungen der Wirbelsäule. II. Über völlstandigen und halbseitigen Mangel von Wirbelkörpern. Virchows Archiv f Path Anat Physiol 278:566–609
150. Feller A, Sternberg H (1929) Zur Kenntnis der Fehlbildungen der Wirbelsäule. I. Die Wirbelkörperspalte und ihre formale Genese. Virchows Archiv f Path Anat Physiol 272:613–640
151. Horn V, Vlach O, Messner P, Silny J (1988) Changes in vertebral end plates in congenital spine deformities. Arch Orthop Trauma Surg 107:231–235
152. Kirmisson E (1910) Scoliose congénitale. Rev d'Orthop 1:21
153. Lehmann-Facius H (1925) Die Keilwirbelbildung bei der Kongenitalen Skoliose. Frankfurter Zeitschrift f Pathologie 31:489
154. Klippel M, Feil A (1912) Un cas d'absence des vertèbres cervicales. Avec cage thoracique remontant jusqu'a la base du crâne. Nouv Iconog Salpétrière 25:223–250
155. Kuhns JG, Hormell RS (1952) Management of congenital scoliosis. Arch Surg 65:250–263
156. Jarcho S, Levin PM (1938) Hereditary malformation of the vertebral bodies. Bull Johns Hopkins Hosp 62:216–226
157. Roberts AP, Conner AN, Tolmie JL, Connor JM (1988) Spondylothoracic and spondylocostal dysostosis. Hereditary forms of spinal deformity. J Bone Joint Surg Br 70:123–126
158. Winter RB, Moe JH, Eilers VE (1968) Congenital scoliosis. A study of 234 patients treated and untreated. Part I. Natural history. J Bone Joint Surg Am 50:1–15
159. McMaster MJ, Ohtsuka K (1982) The natural history of congenital scoliosis: a study of two hundred and fifty-one patients. J Bone Joint Surg Am 64:1128–1147
160. Tsou PM, Yau A, Hodgson AR (1980) Embryogenesis and prenatal development of congenital vertebral anomalies and their classification. Clin Orthop Rel Res 52:211–231
161. Turnpenny PD, Offiah A, Giampietro P, Alman B, Cornier A, Kusumi K, Dunwoodie S, Wade A, Pourquie O (2010) New classification system for segmentation defects of the vertebrae, and pilot validation study. J Bone Joint Surg Br 92(Suppl III):428–9
162. Offiah A, Alman B, Cornier AS, Giampietro PF, Tassy O, Wade A, Turnpenny PD (2010) Pilot assessment of a radiologic classification system for segmentation defects of the vertebrae. Am J Med Genet A 152A:1357–1371
163. Winter RB, Moe JH, Wang JK (1973) Congenital kyphosis: its natural history and treatment as observed in a study of one hundred and thirty patients. J Bone Joint Surg Am 55:223–256

164. McMaster MJ, Singh H (1999) Natural history of congenital kyphosis and kyphoscoliosis. A study of one hundred and twelve patients. J Bone Joint Surg Am 81:1367–1383
165. Tsou PM (1977) Embryology of congenital kyphosis. Clin Orthop Rel Res 128:18–25
166. Shands AR Jr, Bunders WD (1956) Congenital deformities of the spine. An analysis of the roentgenograms of 700 children. Bull Hosp Joint Dis 17:110–133
167. McMaster MJ, David CV (1986) Hemivertebra as a cause of scoliosis: a study of 104 patients. J Bone Joint Surg Br 68:588–95
168. Touzet P, Rigault P, Padovani JP (1979) Les hémi-vertèbres. Classification, histoire naturelle et éléments de pronostic. Rev Chir Orthop 65:173–186
169. McMaster MJ, McMaster ME (2013) Prognosis for congenital scoliosis due to failure of vertebral segmentation. J Bone Joint Surg Am 95:972–979
170. Nasca RJ, Stilling FH 3rd, Stell HH (1975) Progression of congenital scoliosis due to hemivertebrae and hemivertebrae with bars. J Bone Joint Surg Am 57:456–466
171. Tsirikos AI, McMaster MJ (2005) Congenital anomalies of the ribs and chest wall associated with congenital deformities of the spine. J Bone Joint Surg Am 87:2523–2536
172. Gassner M (1982) Kostovertebrale dysplasie. Schweiz Med Wochenschr 112:791–797
173. Cornier AS, Ramirez N, Arroyo S, Acevedo J, Garcia L, Carlo S, Korf B (2004) Phenotype characterization and natural history of spondylothoracic dysplasia syndrome: a series of 27 new cases. Am J Med Genet A 128A:120–126
174. Teli M, Hosalkar H, Gill I, Noordeen H (2004) Spondylocostal dysostosis: thirteen new cases treated by conservative and surgical means. Spine 29:1447–1451
175. Campbell RM Jr, Smith MD, Mayes TC, Mangos JA, Willey-Courand DB, Kose N, Pinero RF, Alder ME, Duong HL, Surber JL (2003) The characteristics of thoracic insufficiency syndrome associated with fused ribs and congenital scoliosis. J Bone Joint Surg Am 85:399–408
176. Campbell RM Jr, Smith MD, Mayes TC, Mangos JA, Willey-Courand DB, Kose N, Pinero RF, Alder ME, Duong HJ, Surber JL (2004) The effect of opening wedge thoracostomy on thoracic insufficiency syndrome associated with fused ribs and congenital scoliosis. J Bone Joint Surg Am 86:1659–1674
177. Ramirez N, Cornier AS, Campbell RM Jr, Carlo S, Arroyo S, Romeu J (2007) Natural history of thoracic insufficiency syndrome: a spondylothoracic perspective. J Bone Joint Surg Am 89:2663–2675
178. Karlin JG, Roth MK, Patil V, Cordell D, Trevino H, Simmons J, Campbell RM, Joshi AP (2014) Management of thoracic insufficiency syndrome in patients with Jarcho-Levin syndrome using VEPTRs (vertical expandable prosthetic titanium ribs). J Bone Joint Surg Am 96 (21), e181. doi:10.2106/JBJS.M.00185
179. Kusumi K, Turnpenny PD (2007) Formation errors of the vertebral column. J Bone Joint Surg Am 89(Suppl 1):64–71
180. Giampietro PF, Dunwoodie SL, Kusumi K, Pourquie O, Tassy O, Offiah AC, Cornier AS, Alman BA, Blank RD, Raggio CL, Glurich I, Turnpenny PD (2009) Progress in the understanding of the genetic etiology of vertebral segmentation disorders in humans. Ann NY Acad Sci 1151:38–67
181. Giampietro PF, Raggio CL, Blank RD, McCarty C, Broeckel U, Pickart MA (2013) Clinical, genetic and environmental factors associated with congenital vertebral malformations. Mol Syndromol 4:94–105
182. Mohamed JY, Faqeih E, Alsiddiky A, Alshammari MJ, Ibrahim NA, Alkuraya FS (2013) Mutations in MEOX1, encoding mesenchyme homeobox 1, cause klippel-Feil anomaly. Am J Hum Genet 92:157–161
183. Lehmann FE (1935) Die Entwicklung von Rückenmark, Spinalganglien, und Wirbelanlagen in Chordalosen Körperregionen von Tritonlarven. Rev Suisse de Zool 42:405–415
184. Kitchen IC (1949) The effects of notochordectomy in Amblystoma mexicanum. J Exp Zool 112:393–415

185. Strudel G (1955) Mécanisme de la genèse des vertèbres chez l'embryon de poulet. Comptes Rendus Acad Sci 240:1725–1726
186. Strudel G (1955) L'influence morphogène du tube nerveux et de la corde sur la différenciation de la colonne vertébrale et de sa musculature chez l'embryon de poulet. Comptes Rendus Soc Biol 149:188–190
187. Goto S, Uhthoff HK (1985) Notochord action on spinal development. Acta Orthop Scand 57:85–90
188. Goto S, Uhthoff HK (1986) Notochord and spinal malformations. Acta Orthop Scand 57:49–153
189. Carey EJ (1922) Direct observations on the transformation of the mesenchyme in the thigh of the pig embryo (sus scrofa), with especial reference to the genesis of the thigh muscles, of the knee- and hip- joints, and of the primary bone of the femur. J Morphol 37:1–77
190. Hueter C (1862) Anatomische studien an den extremitatengelenken neugeborener und erwachsener. Archiv Path Anat Physiol für Klin Med 25:572–599
191. Volkmann R (1862) Chirurgische erfahrungen über knochenverbiegungen und knochenwachsthum. Archiv Path Anat Physiol für Klin Med 24:512–541

Zeitfracht Medien GmbH
Ferdinand-Jühlke-Straße 7
99095 Erfurt, Deutschland
produktsicherheit@kolibri360.de